T0133018

The Rise of the
SUPERCONDUCTORS

The Rise of the
SUPERCONDUCTORS

P.J. Ford
G.A. Saunders

CRC PRESS

Boca Raton   London   New York   Washington, D.C.

Library of Congress Cataloging-in-Publication Data

Ford, P.J. (Peter John)
    The rise of the superconductors / P.J. Ford and G.A. Saunders.
      p. cm.
    Includes bibliographical references and index.
    ISBN 0-7484-0772-3 (alk. paper)
    1. Superconductivity. I. Saunders, G. A. II. Title.

QC611.92.F65 2004
537.6'23--dc22   2004051920

This book contains information obtained from authentic and highly regarded sources. Reprinted material is quoted with permission, and sources are indicated. A wide variety of references are listed. Reasonable efforts have been made to publish reliable data and information, but the author and the publisher cannot assume responsibility for the validity of all materials or for the consequences of their use.

Neither this book nor any part may be reproduced or transmitted in any form or by any means, electronic or mechanical, including photocopying, microfilming, and recording, or by any information storage or retrieval system, without prior permission in writing from the publisher.

The consent of CRC Pressdoes not extend to copying for general distribution, for promotion, for creating new works, or for resale. Specific permission must be obtained in writing from CRC Press for such copying.

Direct all inquiries to CRC Press, 2000 N.W. Corporate Blvd., Boca Raton, Florida 33431.

Trademark Notice: Product or corporate names may be trademarks or registered trademarks, and are used only for identification and explanation, without intent to infringe.

Visit the CRC Press Web site at www.crcpress.com

© 2005 by CRC Press

No claim to original U.S. Government works
International Standard Book Number 0-7484-0772-3
Library of Congress Card Number 2004051920
Printed in the United States of America 1 2 3 4 5 6 7 8 9 0
Printed on acid-free paper

Contents

Contents

Acknowledgments

We would like to thank the Institute of Physics Publishing Ltd in Bristol, who produce *Physics World*, for permission to reproduce from that journal Figures 8.4, 8.6, 9.14, 11.1 and also the four short quotations given on pages 186 and 187. We would also like to thank Mika Seppa of the Brain Research Unit, Low Temperature Laboratory, Helsinki University of Technology, for permission to use Figure 9.14, an image of an arrangement of SQUID sensors for a magnetoencephalography "helmet" appearing on the front cover of the book. Dr George Scott, Professor Michael Springford and The Royal Society of London have allowed us to reproduce Figure 5.8, the Fermi surface of niobium, which appeared in the *Proceedings of the Royal Society* in 1970. Professor Warren Pickett and the American Association for the Advancement of Science have given permission for us to reproduce Figure 5.9, the Fermi surface of YBCO, taken from the journal *Science* in 1992. Kluwer–Plenum press and Professor Heinz Chaloupka have allowed us to reproduce a diagram of an HTS-Meander antenna shown in Figure 9.7, which comes from the *Journal of Superconductivity* in 1992.

The National Academy Press has given us permission to quote from the book *"Physics Through the 1990s: Condensed Matter Physics"* which appears on page 31. The World Scientific Publishing and Professor Ivor Giaver have allowed us to reproduce the quote given on page 101, taken from his 1973 Nobel prize lecture, and appearing in *"Nobel Lectures in Physics 1971–80"*. The Oxford University Press has given permission to use a quote on page 188 taken from a Friday Evening Discourse given by Sir Harold Kroto, which appeared in Volume **67** of the *Proceedings of the Royal Institution*. The American Physical Society has given us permission to reproduce a short quote on page 42 taken from the article "Superconductivity at 93K in a New Mixed Phase Y–Ba–Cu–O Compound System at Ambient Pressure" by Wu et al, which appeared in *Physical Review Letters*, volume **58** in 1987. The publishers Elsevier have allowed us to reproduce on page 30 a short quote from an article on "Excitonic Superconductivity" written by John Bardeen, which appeared in the *Journal of Less Common Metals* in 1978.

Finally we would like to express our thanks and appreciation to several staff members of the Taylor & Francis Group of publishers, especially CRC Press LLC, who have worked extremely hard and been most helpful and cooperative in the production of this book.

Prologue

Ever since their inception in 1901, the Nobel Prizes awarded in the sciences have carried enormous prestige. For the recipients, the award represents ultimate recognition and accolade for their work, and in the eyes of the general public it is the pinnacle of scientific achievement. Often the status of science in a nation is judged in terms of the number of Nobel Prizes received by its citizens. Many Prize winning studies have wide ranging consequences in science and society. The impact on the scientific community of that awarded for physics in 1987 was quite remarkable. The two recipients were J. Georg Bednorz and K. Alex Müller both of whom worked at the IBM Zurich Research Laboratory at Rüschlikon in Switzerland. Their prize citation was *"for their important breakthrough in the discovery of superconductivity in ceramic materials"*. At the time of their award, Bednorz was 37 years old while Müller had turned 60. Why was their work so significant and what was its impact? After all, a Dutch physicist Heike Kamerlingh Onnes, an earlier winner of the Nobel Prize in 1913, had discovered superconductivity long before: shortly after the beginning of the twentieth century.

Many people have heard of *superconductivity* and perhaps are aware that it is a "resistanceless flow of electrical current" and that its technological applications are on the increase. One of the aims of this book is to show that superconductivity is a much richer and more subtle effect than this bald statement implies and that it may well form the basis of a technological revolution in the early part of this millennium.

1986 was the seventy-fifth anniversary of the discovery of superconductivity. What celebrations there were to mark the event were somewhat subdued. Although an enormous amount had been found out about it, superconductivity remained an effect that apparently occurred only at very low temperatures, namely a few degrees above the absolute zero of temperature measured in degrees Kelvin (K). Superconductivity could only be produced under strictly controlled laboratory conditions that precluded widespread applications for it in everyday life. During the 75 years that superconductivity had been known, the critical temperature (T_c) for its onset had only risen slowly from 4.2K, found by Kamerlingh Onnes for the element mercury (Hg), to 23.2K; i.e., 23.2 degrees above the absolute zero of temperature. This highest temperature for T_c was achieved in a thin film of Nb_3Ge, an intermetallic compound of the elements germanium (Ge) and niobium (Nb). Superconductivity in this compound was discovered in 1973 and no higher superconducting transition temperature had been found between then and early 1986. Further gloom was cast by some theoretical work suggesting that the highest temperature at which superconductivity could be expected to occur was unlikely ever to exceed 25K.

All this changed in April 1986, when Bednorz and Müller submitted an article to the journal *Zeitschrift für Physik* in which they described electrical resistance measurements on a complex, ceramic metallic oxide material containing the elements lanthanum (La) – barium (Ba) – copper (Cu) and oxygen (O). They suggested that their data showed evidence for superconductivity above 30K, a substantial increase over the previous record. Their paper appeared in September 1986. It had an almost immediate impact. By December of that year, groups in America and Japan had confirmed and extended the original work. Bednorz and Müller had also carried out further measurements themselves, which greatly strengthened their earlier claims.

There then began one of the most amazing and unprecedented sequence of events in physics: many scientists across the world temporarily abandoned their on-going lines of research in a frantic effort to study these new superconducting ceramics and search for variants with even higher transition temperatures. For a few months, early in 1987, it seemed as if no week went by without the announcement of a further increase in the transition temperature T_c. New journals were created and some existing ones introduced special sections to cope with the flood of papers. Uniquely, daily newspapers added to the hype. Fax and phone were often used to communicate new results. The subsequent Nobel Prize for the work of Bednorz and Müller paid tribute to those exciting events: the citation commented *"This discovery is quite recent – less than two years old – but it has already stimulated research and development throughout the world to an unprecedented extent"*.

At the beginning of March 1987, a paper appeared in the prestigious journal *Physical Review Letters* from two groups working together in the United States of America announcing the discovery of a superconductor with a transition temperature for the onset of superconductivity of 93K. This finding had also been made in a ceramic metal oxide, which in this case contained the elements yttrium (Y) – barium (Ba) – copper (Cu) and oxygen (O). Interestingly, a press announcement was made almost simultaneously in China for superconductivity occurring at around the same temperature. The discovery of this new superconductor created enormous excite-ment because its transition temperature for the onset of superconductivity was above the normal boiling point of liquid nitrogen (77K). Existence of superconductivity in this much higher temperature range greatly enhanced the prospects for widespread applications: an industry based on liquid nitrogen as a refrigerant would be consider-ably cheaper and easier to operate than one based on the previously known super-conductors, which required working with expensive and difficult helium technology.

This second major breakthrough within the space of a few months had important implications for the future direction of several branches of engineering and technology. In turn, this impacted upon both the business and financial communities. The subsequent creation of a number of small and medium sized companies supply-ing the needs of rapidly expanding markets, which exploit these new super-conductors, is an important consequence of the fundamental discovery of Bednorz and Müller. The rapid award of a Nobel Prize in the year following their discovery truly reflected the significance of what they had achieved and capped a momentous period for them. It was also a rare example of abiding by the exact wishes of Alfred Nobel's Will that Prizes should be given to those who *"during the past year*, shall have conferred the greatest benefit on mankind"*.

Superconductivity is an important branch of solid state or condensed matter physics. Sometimes regarded as less exciting than elementary particles physics and cosmology, solid state physics attracts less attention from the media or authors of popular science books. Yet it is no poor relation. It has had an enormous impact on our day to day lives. Many people tend to take for granted the existence of personal computers, color television, video and cam recorders, mobile telephones and text messages, e-mail, the Internet, and a whole host of other seemingly indispensable items of modern living. These all rely extensively for their operation on the spectacular advances in applications of condensed matter physics made over the last fifty years or so. Today, we may be on the threshold of a further important technological revolution based on the new high temperature superconductors.

Before placing the high temperature superconductors in context, we need to summarize the many important discoveries and insights made earlier in superconductivity. Advances could only be made after the development of the science and techniques that enabled temperatures close to the absolute zero to be attained. So, it is first necessary to describe how ever lower temperatures were reached. Then it is possible to recount the dramatic story of the initial discovery of superconductivity by Kamerlingh Onnes at the beginning of the last century. There are interesting parallels between this earlier discovery and that made by Bednorz and Müller.

1   The Long Road to the Discovery of Superconductivity

July 10, 1908, is a largely forgotten landmark in the history of science. It was on that day at the University of Leiden in Holland that Heike Kamerlingh Onnes and his assistants became the first people to liquefy helium (He). In so doing, they paved the way for the study of matter very close to the absolute zero of temperature. Superconductivity, which Kamerlingh Onnes found a few years later, is the most unexpected and exciting of the many discoveries that have taken place as a result of the production of temperatures close to the absolute zero. To appreciate fully the triumph achieved by Kamerlingh Onnes in liquefying helium gas, it is necessary to view his work within its historical context as the end of the long quest to achieve lower and lower temperatures using gas liquefaction. Helium was the last of the gaseous elements to be liquefied. Low temperatures and gas liquefaction are closely interconnected. Even lower temperatures resulted from trying to liquefy more and more gases and the work of Kamerlingh Onnes concluded a history of gas liquefaction which extended for well over 100 years starting from the end of the eighteenth century.

Lowering Temperature in Quest for Its Absolute Zero

The concept of temperature has been dimly perceived since the dawn of man although it is only since the seventeenth century that attempts have been made to define and measure it. There are several candidates for the inventor of the thermometer. By the first half of the eighteenth century, it was realised that for a thermometer to be scientifically useful there have to be defined two fixed points, customarily taken as the freezing and boiling points of water. Fixed-point thermometers are associated with the names of Daniel Fahrenheit, whose scientific work was carried out in the Netherlands, Anders Celsius from Sweden, inventor of the Centigrade scale, and the Frenchman René-Antoine Réaumur. That Fahrenheit chose 32, as the freezing point of water, was tacit recognition that temperatures lower than this could occur but probably only extending down to zero on his scale. Such cold temperatures would have occurred periodically in the Low Countries of Europe in Fahrenheit's day and he may well have thought that his zero point was the lowest temperature that it was possible to reach.

Towards the end of the eighteenth century scientists had become interested in whether all matter existed in the form of solids, liquids and gases. By the 1790s, the Dutch scientist Martinus van Marum was investigating the validity of Boyle's law for the gas ammonia. Boyle's law states that at a fixed temperature the product of the pressure and the volume of a gas is a constant. In those days, it was not clear how universal was this law. Van Marum found that Boyle's law seemed to hold well for ammonia up to a pressure of some seven atmospheres whereupon there was a dramatic reduction in the volume. He had succeeded in liquefying ammonia by the application of pressure. This led to the realization that certain gases at least could be liquefied in this way.

In 1810, the great British scientist Humphry Davy, working at the Royal Institution in London, showed that chlorine gas was an element. In 1823, following a suggestion by Davy, Michael Faraday heated a compound of chlorine in a sealed glass tube and observed at the cold end of the tube droplets of an oily liquid which he rapidly identified as liquid chlorine. Faraday realized that heating the compound in a sealed tube produced a high pressure but that the liquid chlorine had only appeared at the cold end of the tube. He concluded that in order to liquefy chlorine, and presumably other gases, he required both a high pressure and a low temperature. Using these conditions he was able to liquefy several common gases such as carbon dioxide, hydrogen chloride, sulphur dioxide and nitrous oxide. Faraday pursued his low temperature investigations in 1826 and then again in 1845, by which time he had developed more sophisticated equipment enabling him to reach temperatures as low as $-110°C$. That was cold enough to allow him to liquefy several other gases and even solidify some of those that he had previously liquefied. However, despite intense efforts, Faraday and others were unable to liquefy three common gases namely oxygen and nitrogen, the two chief constituents of air, and hydrogen. They became known as the "permanent gases", apparently resisting the notion that all matter could exist in the form of solids, liquids and gases. However, Faraday suspected that it was likely that he had not reached a sufficiently low temperature to liquefy these particular three gases.

It was not until 1869, with the completion of brilliant experiments on carbon dioxide carried out by Thomas Andrews at Queens University Belfast, that the idea became firmly established of a critical temperature above which no amount of pressure will reduce a gas to a liquid. Andrews spent several years studying the variation of the volume with pressure of a fixed mass of carbon dioxide at different temperatures to produce the pressure–volume isotherms. From these curves, it was possible to study in detail the nature of the gas–liquid transition; and from this, the concept of a critical temperature became clear. Andrews' experiments were placed on a firm theoretical footing by the work of the Dutch scientist Johannes Diederik van der Waals. Boyle's law is only true for ideal gases. In 1872, van der Waals considered the case of real gases for which the finite volume of the gas molecules and also the interaction between molecules has to be considered. His work gave a remarkably good qualitative description of the behavior of the gas–liquid transition and led to his award of the Nobel Prize for physics in 1910.

The development of the science of low temperatures continued when both oxygen and nitrogen were finally liquefied at the end of 1877 almost simultaneously by Louis Cailletet in France and Raoul Pictet in Switzerland. They employed very different methods. Cailletet observed that when oxygen or nitrogen (already at a low temperature) are compressed to a very high pressure, which is then suddenly released, droplets are formed which could be identified as being liquid oxygen or nitrogen. Such a sudden increase in volume does not allow time for heat either to enter or leave the system; an expansion of this kind is said to be adiabatic. When a rapid expansion occurs like this, there is no time for the system to change its disorder; hence, an adiabatic process is one which takes place at constant entropy. Cailletet built various versions of his expansion machine and encouraged others to experiment with them. One went to the Jagiellonian University in Cracow, which was then part of the Austro Hungarian empire. Two scientists, Zygmunt Wrobleski and Karol Olsweski, managed to make modifications to Cailletet's apparatus and produce small quantities of both liquid oxygen and nitrogen instead of the droplets obtained by Cailletet. Oxygen was found to boil at $-183°C$ and nitrogen at $-196°C$. By the time that these gases were liquefied the Kelvin scale of temperature, based on thermodynamic principles, had become well established. There exists an absolute zero of temperature at $-273°C$ below which it is impossible to go. This is defined as zero degrees Kelvin (K). On the Kelvin temperature scale, oxygen boils at 90K and nitrogen at 77K; this scale is used throughout the remainder of the book.

Obtaining lower temperatures was helped by Pictet who used a "cascade" technique requiring a series of gases with progressively lower values of the critical temperature. That is what made it possible to reach a temperature below the critical temperature of oxygen and nitrogen, which could then be liquefied by applying pressure. Cailletet and Pictet were jointly awarded the Royal Society Davy medal in 1878 for their work. Their achievements were significant in that they destroyed the concept of the "permanent gas" and suggested that hydrogen could also be liquefied. They had opened up a new lower temperature region in which basic investigations of the electrical resistivity and other properties could be made. Their work also paved the way for the later development of a liquid-gas industry, which today is a multi-million pound operation with widespread applications.

Another of Cailletet's expansion engines arrived in England and was used by James Dewar at the Royal Institution. Dewar was said to be short both in stature and in temper but at the same time was a superb experimentalist. Using the expansion engine, he produced the first liquid oxygen and nitrogen in the United Kingdom. In order to prevent the cold liquids from evaporating, he invented vacuum insulated vessels, which are presently referred to as dewars. Anyone enjoying a hot cup of tea on top of a mountain or picnicking on the beach owes a debt of gratitude to Sir James. In 1898, he became the first person to liquefy hydrogen, which boils at just over 20K at atmospheric pressure. Dewar's technique involved a cooling process that depends on the Joule–Kelvin effect: when a gas under high pressure is expanded through a nozzle, it can cool. His system relied for its success on the use of an efficient heat exchanger so that the outgoing gas from the nozzle is used to cool the incoming high pressure gas. Playing with liquid hydrogen proved a most hazardous occupation: two of Dewar's assistants each lost an eye. By pumping on the liquid, thereby reducing its vapor pressure, he obtained solid hydrogen at 13K. To reach a temperature which was only 13K above the absolute zero of temperature was a major technological achievement, which was justly recognized. At that time, he thought that hydrogen would be the last gas to be liquefied. However, a few years earlier helium had been discovered as a new element and it turned out that this had an even lower boiling point.

More Discoveries: the Noble Gases

Around the turn of the twentieth century, many fundamental discoveries were made which transformed completely our view of physics. In 1895, Röntgen discovered X-rays; in 1896, Becquerel and the Curies discovered radioactivity; in 1897, J.J. Thomson discovered the electron; in 1900, Planck introduced the concept of the quantized nature of energy in order to explain the radiation emitted from a black body; and in 1905, Einstein put forward his ideas on special relativity and also explained the photoelectric effect in terms of quantized particles of light known as photons. These discoveries were widely acclaimed, had great significance, and each scientist obtained a Nobel Prize in physics. Taken together, they marked the beginning of a new era, the dividing line between classical and modern physics.

It was also in the 1890s that a new series of elements known as the rare or noble gases was discovered. Working in his private laboratory at Terling in Essex, the eminent scientist John William Strutt, the third Baron Rayleigh, made some highly accurate determinations of the density of hydrogen and oxygen. In 1892, he examined nitrogen and observed a small difference in his measured values of the density of nitrogen obtained from the air and from other sources such as ammonia. With great tenacity, he pursued the source of this discrepancy and concluded that the nitrogen from the air was contaminated in some manner by a heavier gas. In the course of his investigations, Rayleigh had discussions and correspondence with Sir William Ramsay who at the time was Professor of Chemistry at University College, London. Rayleigh encouraged Ramsay to study this problem and working largely independently, they both demonstrated that the nitrogen from the air contained a new element, which was named argon from a Greek word meaning lazy.

Rayleigh and Ramsay presented their results in a joint paper to the Royal Society; this publication was one of the few in which Rayleigh had a coauthor. The chemically inert nature of argon gas suggested that it had a valency of zero and it was believed that it was one of a new group of elements in the Periodic Table. The following year, Ramsay used a spectroscope to examine the gases emitted on heating the mineral pitchblende and observed a bright yellow line, which he attributed to yet another unknown element. A similar yellow line had already been observed in 1869 by the French astronomer Pierre Jansen who had first used a spectroscope to examine the light spectrum of the sun. It was concluded that the line was due to a new element which was later called helium after the Greek word *helios* for the sun. Ramsay tried unsuccessfully to find evidence of other inert gaseous elements in terrestrial minerals. In 1898, together with his assistant Maurice Travers, Ramsay embarked on an exhaustive series of experiments in which they carried out a fractional distillation of large quantities of argon and examined spectroscopically the gases that had been evaporated off. In so doing, they discovered three more inert gaseous elements, which were called neon, xenon and krypton. Ramsay was awarded the 1904 Nobel Prize in chemistry for his discovery of this new series of inert noble gases, including helium, while that same year Rayleigh was awarded the prize for physics mainly for his work on the density of gases in the air and the discovery of argon.

The properties of helium gas were studied intensively with a view to its lique-faction. Dewar concluded from experiments that it probably had a critical temperature between 5 and 6K. He had a major problem with obtaining sufficient quantities of pure helium gas. The gas he used was obtained from the mineral springs in Bath and this was contaminated with neon which on freezing kept on blocking the pipes and valves of his apparatus. It is a great pity that Dewar and Ramsay did not collaborate since their combined talents and expertise was such that together, they might well have liquefied helium. Unfortunately the two men were not on speaking terms as a result of a bitter dispute some years earlier over priority for the first liquefaction of hydrogen. Dewar failed in his attempts to liquefy helium and he rapidly lost interest in low temperatures after the Dutch scientist Kamerlingh Onnes succeeded.

Kamerlingh Onnes and the Liquefaction of Helium

In January 1807, a barge carrying gunpowder was moored in the centre of the city of Leiden in Holland. It exploded – resulting in the destruction of some five hundred houses. This unconventional but effective means of land clearance allowed the university in the city to expand and the Leiden Physics Laboratory came into existence. Today this laboratory is inextricably linked with the name of Heike Kamerlingh Onnes. He was born in the town of Groningen in 1853 and came from a wealthy family of merchants and manufacturers. Onnes went to school in Groningen before studying at its university, obtaining a doctorate in 1879. For a few years he taught physics in Delft before being appointed to the chair of physics at Leiden in 1882, a post that he was to hold for 42 years. Kamerlingh Onnes gave a famous inaugural lecture in which he set out his philosophy of doing science. This was encapsulated in a motto "*door meten tot weten*" – through measurement to knowledge – which was inscribed above his laboratory door.

Over a period of many years Kamerlingh Onnes, a fatherly but autocratic figure, systematically built up the low temperature facilities at Leiden. In 1892, he was able to produce several litres an hour of liquid air or liquid oxygen; by 1906, he had a hydrogen liquefier working that was capable of delivering four litres of liquid per hour. These meticulous preparations were to enable him to carry out the assault on the liquefaction of helium. He had already obtained an excellent source of very pure helium from monazite sand. This was probably decisive in enabling him, rather than Dewar, to be successful in liquefying helium.

The successful experiment occurred on July 10, 1908. It was an exhausting day, work beginning at 5a.m. By 7p.m., the thermometers, which were used, were no longer showing a change in temperature and it appeared that the attempt had failed. It was at this point that a visitor from the Chemistry department, who had stopped by to see how the famous experiment was progressing, suggested that the cryostat should be illuminated from underneath and this showed that already about a litre of liquid helium had collected. Failure was transformed instantly into triumph! The difficulty in observing the helium is due to it being a colorless, transparent liquid with a refractive index close to one: light does not bend appreciably at its surface. In addition, the surface of the liquid is almost horizontal – it has a very flat meniscus: liquid helium has a low surface tension. Kamerlingh Onnes used his pumps to reduce the vapour pressure above the liquid, producing a lowering of the temperature to around one degree above the absolute zero of temperature. He noted that the liquid did not solidify even at that low temperature. However, it seems that he did not observe that the liquid changes its behavior at a temperature of around 2.2K. It is now realized that above this temperature, the liquid helium is violently agitating while below, it suddenly becomes quiescent. This effect is associated with a change to a most remarkable and unique superfluid state at 2.19K, below which quantum mechanical effects dominate the behavior of the liquid. There are striking analogies between superfluidity in liquid helium and superconductivity; the former is a frictionless flow of atoms and the latter a frictionless flow of electrons.

Kamerlingh Onnes made the point that in Leiden, he had achieved the lowest temperature ever reached on Earth. He was not to know that in fact he had reached a lower temperature than that (3K) of the background radiation of the universe itself. In a real sense he had outstripped nature. He can be regarded as one of the first modern scientists relying on large-scale facilities, excellent technical support and funding to match in order to reach their goal. He established an important school of instrument makers and glassblowers at Leiden so as to train people to the high level needed to provide technical support for his research. He was fortunate in having an outstanding laboratory chief technician, Gerrit Jan Flim, whose expertise and skills played an invaluable role in this first successful liquefaction of helium and subsequently in the discovery of superconductivity. Kamerlingh Onnes contrasts with people like Sir James Dewar, Sir William Ramsay and Lord Rayleigh who often worked alone, apart from one or two laboratory assistants, and who can be regarded as representing the peak of the classical era of physics.

As for all gases, in order for helium to condense into a liquid, the interaction between the helium atoms must be greater than the effect of thermal agitation. The reason why helium is so difficult to liquefy is due to the weakness of the interaction between its atoms, which reflects its inert nature. This in turn depends on the electronic structure of its atoms. As a result of some elegant experiments involving the scattering of alpha particles by thin foils of metallic gold, a planetary model of the atom was established. Alpha particles are helium nuclei and Ernest Rutherford and his colleagues used them to brilliant effect in understanding the structure of the atom. Hans Geiger and Ernest Marsden carried out the experiments in 1909 in Rutherford's laboratory at the University of Manchester. They observed occasional large angle scattering of the alpha particles suggestive of a heavy positive nucleus in a nearly empty atom rather than the positive charge being uniformly distributed in the atom as proposed in a model put forward by J.J. Thomson. In the Rutherford model of the atom, the negatively charged electrons orbit around a positively charged nucleus. The different elements correspond to the different values for the positively charged nucleus and the same number of electrons orbiting around in order to produce an electrically neutral atom. Hydrogen is the simplest of the elements with one proton in the positively charged nucleus and one electron orbiting around it. Helium is the next element with two protons and two neutrons in its nucleus around which orbit two electrons. The two electrons form a tightly bound closed shell (labeled K) and because this shell is closed, it has its full complement of electrons. This is why helium atoms are inert and do not interact with other atoms to form compounds by exchanging electrons. A very weak interaction between helium atoms takes place via the so-called van der Waals interaction and it is only at very low temperatures that this interaction is sufficiently strong enough to overcome the disruptive effect of thermal agitation to allow helium to condense into a liquid.

The Electrical Resistance of Metals and the Discovery of Superconductivity

The production of sufficient quantities of oxygen, nitrogen and liquid air led to the first studies of matter at low temperatures. Dewar played a prominent role in the early work. Between 1892 and 1895, he collaborated with John Ambrose Fleming, a professor of Electrical Engineering at University College, London, in a comprehensive investigation of the electrical resistance of pure metals between +200°C and −200°C, close to the lowest temperature then available. Their measurements of the electrical resistance as a function of temperature suggested that the resistance might vanish at the absolute zero of temperature. Fleming later became famous with his invention of the thermionic valve in 1904, so beginning electronics.

After Kamerlingh Onnes had first liquefied helium, Dewar lost interest in low temperatures, and this meant that until the early 1920s, Kamerlingh Onnes's laboratory at Leiden was the only place where extensive research at very low temperatures was pursued. Prominent among this research was the investigation into the electrical resistance of pure metals. Almost from its discovery, the electron was regarded as the particle responsible for electrical conduction and as early as 1900, the German physicist Paul Drude produced a simple theory for the electrical conductivity of pure metals. The arrangement of the elements in the Periodic Table reflects the number of electrons needed to neutralize the positive charge on the nucleus and the way in which these electrons occupy a series of shells around the nucleus. It was later to be found that the number of electrons in each shell is determined by the Pauli exclusion principle which states that no two electrons in an atom can have the same set of quantum numbers. In his model, Drude assumed that electrons from the outermost shell detach themselves from the atoms to become conduction electrons. These migrate in an applied electric field to give rise to an electric current. Loss of the moving conduction electrons causes the atoms to

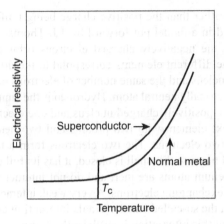

Figure 1.1 Comparison between the electrical resistivity of a superconducting and a normal metal at low temperatures (less than 20K). For a metal which goes superconducting there is a sudden, sharp drop to zero resistivity, which takes place at the superconducting transition temperature T_c. In contrast for the normal metal the electrical resistivity eventually becomes flat as the temperature is decreased, that is, it has no temperature dependence. The low temperature value is known as the "residual resistivity" and results from the current carrying electrons being scattered by mechanisms that do not alter with temperature. The increasing resistivity in the normal state at higher temperatures is due to the scattering of electrons by the lattice vibrations (see Chapter 5).

become positively charged and form positive ions. These ions vibrate about their mean positions within the lattice and scatter the conduction electrons – giving rise to electrical resistance. As the temperature is lowered, the amplitude of the ionic vibrations becomes smaller and so the passage of the conduction electrons through the lattice becomes easier – resulting in a decrease of the resistance.

An interesting theory was put forward in 1902 by Lord Kelvin, the great Victorian scientist, who was responsible, among many other things, for the laying of the first transatlantic cable and after whom the absolute scale of temperature is named. Kelvin suggested that at very low temperatures, the conduction electrons would reattach themselves to the atoms and as a consequence would no longer be able to move freely, resulting in the electrical resistance passing through a minimum before rising rapidly. It is believed that a plan to test this idea led Kamerlingh Onnes to pursue an investigation of the electrical resistance of pure metals. Measurements, which were the first made down to liquid helium temperatures, showed that the resistance behavior of nominally pure gold and platinum decreased linearly with temperature before flattening off below about 10K as shown in Figure 1.1. This flattening off was attributed to the presence of traces of impurities in the metal, as well as to the metallurgical state of strain in the wire, which we now know produces an important type of defect called a dislocation.

In those days, the element mercury, in particular, was able to be produced to an extremely high level of purity: since it is a liquid at room temperature, it is possible to redistil it many times to produce a very pure form. Measurements made on mercury, which solidifies at $-38.8°C$, taken successively in the temperature range

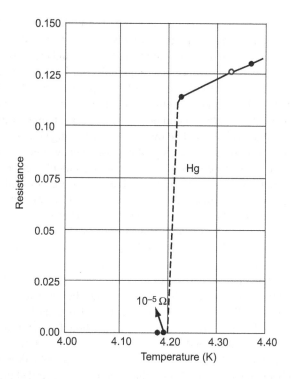

Figure 1.2   The electrical resistance in ohms of mercury (Hg) in the vicinity of 4.2K, indicating a sharp drop at the superconducting transition. (After Kamerlingh Onnes.)

of liquid oxygen, liquid nitrogen and liquid hydrogen all showed the expected decrease with temperature of the electrical resistance. However, at the temperature of liquid helium, the resistance appeared to be zero, which Kamerlingh Onnes attributed at first to an electrical short. Repetition of the experiment produced the same result, as did using a different configuration of the mercury. Tracing the source of this "short" proved very difficult. It was only when a laboratory assistant, who was supposed to ensure that the temperature of the cryostat did not rise above the normal boiling point of liquid helium of 4.2K, dozed off during an experiment, thereby allowing the temperature to rise slowly, did the resistance suddenly re-appear. By such fortunate human foibles can great discoveries be made! This sudden reappearance of the electrical resistance meant that there was in fact no "short". What was being observed was an exciting and totally unexpected new effect – superconductivity.

The individual who carried out the decisive measurements on mercury was Gilles Holst, a doctoral student of Kamerlingh Onnes, and he should probably be given much greater credit for the discovery of superconductivity than history has given him to date. Holst later had a distinguished scientific career becoming the first director of the prestigious Philips research laboratories at Eindhoven. The original measurements of the electrical resistance of mercury, reproduced in Figure 1.2, indicate the sharp decrease near 4.2K that first showed the onset of superconductivity.

It was a quirk of nature that in the case of mercury, the transition to the super-conducting state occurred at a temperature that was very close to that of the normal boiling point of liquid helium of 4.2K. Kamerlingh Onnes and his assistants agonized for a long time as to whether or not the strange behavior being observed was some curious effect associated with the onset of the production of liquid helium. Confirmation that the findings represented something new and very exciting, came when similar behaviour was observed in both lead and tin at temperatures of 7.2 and 3.7K respectively, that is above and below the normal boiling point of liquid helium. Kamerlingh Onnes carried out some elegant experiments designed to show that the electrical resistance in the superconducting state was genuinely zero rather than just extremely low. Strictly speaking, it is not possible to measure a quantity that is zero: an experiment can only establish that the quantity must be smaller than a certain limit set by the sensitivity of the measuring equipment. These experiments consisted of inducing a current into a superconducting loop and observing any decay in the magnitude of the current, which was generally seen as a reduction in the associated magnetic field. He observed no diminution in the current, which as a result became known as a "persistent current". After an intensive period of research, Kamerlingh Onnes was eventually able to convince himself that the electrical resistance was indeed zero within the limits of his experiments and that he had stumbled on a strange and inexplicable new effect. His work on the liquefaction of helium and the subsequent discovery of superconductivity had opened up wondrously exciting new fields in science and its applications. For his work on the liquefaction of helium and the discovery of superconductivity Kamerlingh Onnes was awarded the Nobel Prize for Physics in 1913.

The strength of Dutch science in the years around the turn of the twentieth century can be judged from the fact that, in addition to the Nobel Prizes in physics won by Kamerlingh Onnes in 1913 and Van der Waals in 1910, Pieter Zeeman and Hendrik Anton Lorentz shared the 1902 prize. The first Nobel Prize winner in chemistry in 1901 was Jacobus van't Hoff for his enunciation of the laws of chemical kinetics and the laws governing the osmotic pressure of solutions. In 1897, while working at Leiden, Zeeman discovered the splitting of spectral lines by a magnetic field while Lorentz carried out important theoretical investigations on the electron and the propagation of light. Lorentz analyzed the experiment of Zeeman and obtained a value for the ratio of the charge to the mass of the electron e/m, which was very close to that found earlier by J.J. Thomson. The Lorentz force, which describes the motion of electrons moving in a magnetic field, appears in the discussion of the use of superconductors to generate high magnetic fields in Chapter 8.

Kamerlingh Onnes immediately appreciated that his new discovery potentially had important technological implications centered in particular on the production of high magnetic fields. It was well known that passage of an electrical current gives rise to a magnetic field. Therefore, a wire can be wound into a coil, called a solenoid, to produce an electromagnet. The disadvantage of using normal metals is that such a solenoid has a finite electrical resistance and passage of a large current necessary to produce a high magnetic field also produces Joule heating which eventually can burn out the solenoid. Clearly, a superconducting solenoid capable of delivering a high current with zero electrical resistance and hence no Joule heating, would be

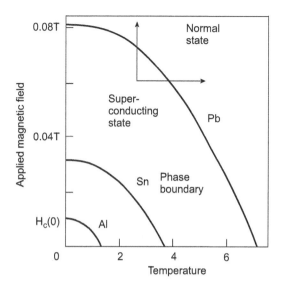

Figure 1.3 The way in which critical magnetic fields of superconducting lead (Pb), tin (Sn) and aluminium (Al) depend upon temperature. The phase boundary (a parabolic curve) defines the transition from the superconducting to the normal state. The normal state is reached if either the temperature rises above a critical value T_c or the magnetic field rises above a critical value H_c. This is shown for lead; when the magnetic field applied to superconducting lead is increased along the direction of the vertical arrow, the metal goes into normal state where the arrow crosses the phase boundary line. The horizontal arrow shows how a superconductor can be driven normal by increasing the temperature.

enormously advantageous. However, Kamerlingh Onnes quickly ran into problems. He observed that passage of a large current produced a magnetic field which destroyed superconductivity. The solenoids ceased to be superconducting in magnetic fields which exceeded a few hundredths of a Tesla. There is a limit, known as the critical current density, to the amount of current that can be carried. In addition, application of a strong enough magnetic field to increase the current flow up to its critical value destroys superconductivity. This critical magnetic field H_c, required to cause the superconductor to function normally, is different for each superconducting metal; examples are shown in Figure 1.3. This figure also illustrates the experimental finding that the critical magnetic field increases with decreasing temperature down to a limit $H_c(0)$ at zero temperature. It is a "phase" diagram, which shows the ranges of temperature and magnetic field over which the superconducting and normal states exist. Increasing either temperature or applied magnetic field beyond critical limits, defined by the boundary line between the superconducting and normal phases, destroys the superconductivity.

Kamerlingh Onnes realized that in order for superconductivity to be of much practical use both the critical temperature T_c for the onset of superconductivity and the critical magnetic field H_c had to be increased dramatically. Much of the later history of experimental superconductivity has revolved around trying to achieve these conditions.

2 Superconductivity Reveals Its Mysteries

New Superconducting Materials Widen the Horizon

Superconductivity remained an intriguing but obscure effect for many years after its discovery. The work at Leiden had established the most spectacular property of a superconductor: its electrical resistance is zero. By contrast a normal metal shows an electrical resistance. It is impossible to measure a zero resistance; however it is now known that a dc current flowing round a superconducting ring remains unchanged for years. This persistence places an upper limit on the electrical resistivity of at least 10^{15} times smaller than that of a normal metal at the same temperature. No voltage is observable across a superconductor when it is included in an otherwise normal (i.e. non-superconducting) circuit and a direct current is flowing: for all practical purposes the resistance is zero.

Until 1923, Leiden remained the only place in the world that possessed liquid helium facilities and hence had a monopoly in the investigation of superconductivity. The steady stream of results flowing from the laboratory was generally published in the *"Communications of the University of Leiden"*. During those early years in Leiden, Onnes showed that, in addition to mercury, the metallic elements tin and lead also become superconducting and established that the superconducting transition is extremely sharp (to within about 1mK). The unique position held by Leiden in that early long period contrasts markedly with the explosion of research activity and dissemination of results that followed the discovery of the high temperature superconductors!

By 1923, two other laboratories had developed low temperature facilities, and both began research into superconductivity. These were the University of Toronto in Canada and the Physikalisch-Technische Reichsanstalt at Berlin in Germany. Within the next ten years, it became apparent that superconductivity is not just confined to a few pure elements: a large number of alloys and chemical compounds show zero resistance. Well over a thousand superconducting materials are now known.

Over the years, the number of elements that can become superconducting has increased, as techniques for reaching lower temperatures or making purer metals have been developed. Figure 2.1 shows the Periodic Table marked with the many superconducting elements known today. Niobium is the element with the highest superconducting transition temperature of 9.5K.

As a rule, the strong magnetic fields that exist in the ferromagnetic elements iron, cobalt and nickel prevent such materials from also being superconducting. When searching for the presence of superconductivity in a material, it is extremely important to eliminate as far as possible all traces of iron and other elements that have the capacity of becoming ferromagnetic. A good case of the influence of such an impurity is shown by molybdenum: when pure, this element has a transition temperature of 0.92K but only a few parts per million of iron prevent it from becoming superconducting.

Interestingly enough, a group from Osaka in Japan, led by Katsuya Shimuzu, has recently demonstrated that in iron itself ferromagnetic effects are instrumental in destroying superconductivity. Normal ferromagnetic iron has a body-centred cubic (bcc) structure. On application of pressures above 10GPa, iron is transformed to a hexagonal close packed (hcp) structure and then is not ferromagnetic; it was predicted in the early 1970s that this phase might become superconducting. A comprehensive investigation by the Japanese group showed that this is indeed the case at temperatures below 2K and pressures between 15 and 30GPa.

The application of high pressure has proved to be a powerful technique for inducing superconductivity and quite a few elements only become superconducting under pressures high enough to cause them to be transformed into a more dense, metallic phase. These include selenium, germanium, arsenic, phosphorus, cerium, barium, yttrium and caesium. Jorg Wittig has discovered many of these superconducting transitions and his investigations in this area have been particularly fruitful. The observation that caesium can be produced in a pressure induced form that goes superconducting at 1.5K is especially important because previously it was widely believed that superconductivity in the alkali elements would not be observed until extremely low temperatures, in the micro-degree range, were reached.

Evidence for superconductivity in another alkali metal has recently been reported for lithium under high pressure by two independent teams in Japan and the U.S. Although the details of the results differ slightly, both groups found that under increasing high pressure lithium transforms to a series of different phases. The transition temperature increased with pressure. The superconducting transition temperature is remarkably high: the Japanese group reported a value of 20K at a pressure of just under 50 GPa. To date more than twenty elements have been found to become superconducting under high pressure adding to nearly thirty more at atmospheric pressure.

For many years there has been speculation that hydrogen might become metallic at sufficiently high temperatures and pressures and also that it might even become superconducting. One way of inducing extremely high pressure is to impact a projectile onto a material. Shock-wave experiments at the Lawrence Livermore National Laboratory in California, carried out in 1996, have suggested that hydrogen undergoes a phase transition to form a liquid metallic state at temperatures around 3000K and pressures of 140GPa. However all claims for superconductivity in this metallic form of hydrogen have been vociferously rejected.

To date, there has been no accepted evidence that superconductivity occurs in either the noble elements copper, silver and gold or the alkali metals sodium and potassium. But remember that the search for superconductivity among the elements

Figure 2.1 (Periodic Table)

1	2	3	4	5	6	7	8	9	10	11	12	13	14	15	16	17	18
Li	**Be** 0.03											B	C	N	O	F	Ne
Na	Mg											**Al** 1.14 / 105	**Si** 6.7	**P** 4.8-6.1	S	Cl	Ar
K	Ca	Sc	**Ti** 0.39 / 100	**V** 5.38 / 1420	Cr	Mn	Fe	Co	Ni	Cu	**Zn** 0.87 / 53	**Ga** 1.09 / 51	**Ge** 5.4	**As** 0.5	**Se** 6.9	Br	Kr
Rb	Sr	**Y** 1.5, 2.7	**Zr** 0.55 / 47	**Nb** 9.50 / 1980	**Mo** 0.92 / 95	**Tc** 7.77 / 1410	**Ru** 0.51 / 70	**Rh** 0.0003 / 0.049	Pd	Ag	**Cd** 0.56 / 30	**In** 3.40 / 293	**Sn** 3.72 / 309	**Sb** 3.6	**Te** 4.5	I	Xe
Cs 1.5	**Ba** 1.8, 5.1	**La** 4.8, 5.9 / 1100	**Hf** 0.12	**Ta** 4.48 / 830	**W** 0.01 / 1.07	**Re** 1.4 / 198	**Os** 0.66 / 65	**Ir** 0.14 / 19	Pt	Au	**Hg** 4.15 / 412	**Tl** 2.39 / 171	**Pb** 7.19 / 803	**Bi** 3.9, 7.2, 8.5	Po	At	Rn
Fr	Ra	**Ac**															

Ce 1.7	Pr	Nd	Pm	Sm	Eu	Gd	Tb	Dy	Ho	Er	Tm	Yb	**Lu** 0.1, 0.7
Th 1.37 / 162	**Pa** 1.3	**U** 0.2	**Np** 0.07	Pu	Am	Cm	Bk	Cf	Es	Fm	Md	No	Lr

Figure 2.1 The Periodic Table showing most elements that are superconducting. Those elements shown in heavy font are superconducting at atmospheric pressure. The critical temperature is given underneath the element symbol. The critical magnetic field, given in Gauss (10^{-4} Tesla) at the absolute zero, is shown in the bottom row of those boxes. Those elements, which only become superconducting under high pressure, are shown in the shaded boxes.

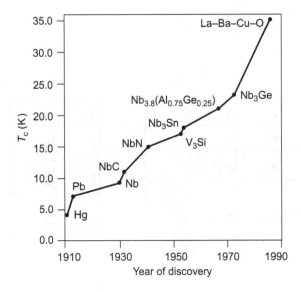

Figure 2.2   A plot of the highest superconducting transition temperature obtained against the year for the period between the discovery of superconductivity in 1911 and the advent of high temperature superconductors in 1986.

continues to surprise us. After all superconductivity has recently been observed in compacted platinum powder (in contrast to the bulk form of the metal) at very low temperatures.

While an element itself may not be superconducting, combinations of it with other elements can be. Combinations that are found to be superconducting include copper–sulphur and gold–bismuth in suitable proportions and form. This is significant because in both these cases (and in many others), neither of the component elements is superconducting at atmospheric pressure; this observation clearly establishes that superconductivity is not associated with specific types of atoms but that their arrangements and combinations play a role.

The 1930s saw considerable advances in understanding the two great topics of low temperature physics: superconductivity and superfluidity in liquid helium. By then, several more laboratories had acquired liquid helium facilities and hence, were able to carry out research into low temperature physics. These included Oxford, Cambridge and Bristol in the United Kingdom, Moscow and Kharkov in Russia, Washington in America and Breslau in Germany. Gradually during the 1930s the superconducting transition temperature attained in compounds and alloys began to increase. By 1941, Eduard Justi and his team at the Physikalisch-Technische Reichs-anstalt in Berlin had achieved a transition temperature of 15K in the compound NbN. The increase in the superconducting transition temperature during the 75 years since the initial discovery in 1911 is shown in Figure 2.2.

A fundamental understanding of the nature of matter on the atomic scale was emerging from the developments in quantum mechanics during the 1920s. The years between 1928 and 1932 saw the successful application of this truly awesome theory

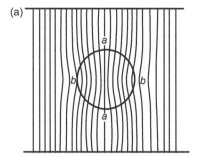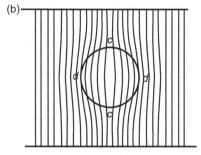

Figure 2.3 Lines of magnetic force (flux lines) around (a) a paramagnetic and (b) a diamagnetic sample. The figure is based on one appearing in an article by Faraday published in his *"Experimental Researches in Electricity"* in 1852.

to a variety of problems in metal physics. It was widely hoped and believed that ideas founded in quantum theory would throw further light on the nature of superconductivity. Indeed that was to happen but not until nearly thirty years later. Despite the efforts of some of the most prominent theorists of the day, including Niels Bohr, Werner Heisenberg and Felix Bloch, little headway was made in explaining superconductivity. As yet, not enough was known about the experimental facts. In the early years, it had been widely believed that superconductivity was solely the effect of the sudden complete loss of electrical resistance. No sudden change at the transition temperature had yet been observed in any physical quantity other than those associated with the conduction electrons such as the specific heat or the thermoelectric power. The next step was to be the discovery of a magnetic property just as fundamental as the zero resistance.

The Meissner Effect, a Fundamental Criterion for Superconductivity

Although it was well known that application of quite a small magnetic field destroys superconductivity (Chapter 1), the actual magnetic effects involved were not clear. In 1933, Walther Meissner and Robert Ochsenfeld, working in Berlin, made a surprising and fundamental discovery in the magnetic behavior of a superconductor. It had been nearly a century earlier that Michael Faraday working at the Royal Institution of Great Britain in London had demonstrated the distinction between paramagnetic and diamagnetic behavior depending on how a material responds to an applied magnetic field. When a material is placed in a uniform magnetic field, if the lines of magnetic flux crowd together into the material it is said to be paramagnetic while if they move further apart it is diamagnetic; this is shown in Figure 2.3 which comes from one of Faraday's original papers. A paramagnetic material is attracted towards an applied magnetic field whereas a diamagnetic material is repelled from it. Faraday's work on paramagnetism and diamagnetism was the first step towards a

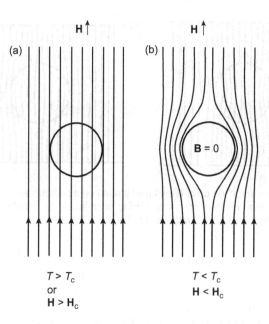

Figure 2.4 The Meissner effect. The magnetic lines of force (flux lines) in and around a metal, which goes superconducting: (a) In the normal state above the critical temperature T_c or when the applied magnetic field is greater than the critical field H_c. (b) When the sample is cooled below T_c, the metal becomes superconducting and the magnetic field is completely expelled from the superconductor; hence the flux density **B** inside is zero.

gradual recognition of the great richness in the magnetic behavior of materials. Before his studies the only widely known form of magnetism had been that of the spontaneous ferromagnetism exhibited in iron (see Chapter 5).

All materials show some type of magnetic behavior but that found in a super-conductor is unique and quite remarkable. Meissner and Ochsenfeld observed that when pure tin is cooled in the presence of a magnetic field, on reaching its super-conducting transition temperature, the magnetic flux is suddenly *completely expelled* from its interior, as shown in Figure 2.4. This discovery later became known as the Meissner effect. Since diamagnetism is a measure of the ability of a material to shield its interior from an applied magnetic field, the complete exclusion of flux means that a superconductor is a *perfect diamagnet*.

It is now recognized that such perfect diamagnetism, in addition to zero resistivity, is a fundamental and equally basic property of the superconducting state. Perfect diamagnetism cannot be explained by starting from zero electrical resistivity (i.e. infinite electrical conductivity) and applying the relevant electromagnetic equations, derived earlier by James Clerk Maxwell, which many scientists consider to be the most far reaching contribution to theoretical physics in the nineteenth century. Radio, television and satellite communications ultimately stem from these equations but that's another story (see Chapter 9)! Application of Maxwell's equations requires that the magnetic flux inside a completely superconducting body cannot change and this in turn would imply that any magnetic flux present inside the superconductor before

Figure 2.5 Levitation, induced by the persistent surface current and the Meissner effect, of a powerful, small samarium–cobalt permanent magnet above a YBCO polycrystalline sample produced by students in the Department of Physics of the University of Bath.

it is cooled into the superconducting state would remain frozen in. Thus, the magnetic flux inside a superconductor would depend upon its magnetic history – whether it had been cooled to below its superconducting transition temperature either with or without the presence of a magnetic field. Such an experimental situation would mean that the superconducting state is metastable rather than being in true thermodynamic equilibrium. However, that is not the case. The confusion was resolved by the discovery of the Meissner effect that the magnetic flux is completely excluded at the superconducting transition temperature; this means that, contrary to earlier wide-spread thinking, a true thermodynamic state in fact does exist. The behavior when a magnetic field is applied to a superconductor is reversible. Raising the temperature above the critical temperature T_c or applying a magnetic field greater than a certain critical value H_c (see Figure 1.3) causes a sample to return to its normal state and flux penetration takes place.

Superconductors in which the flux exclusion is complete for any applied magnetic field less than the critical value are known as type I. Above this critical value, the material goes normal and flux penetration is total. These type I superconductors are usually elements or simple alloys.

Demonstrating the Meissner effect to a general audience used to be no easy task. Lead, which has a critical temperature T_c of 7.2K and can be readily fabricated into a dish, was the most convenient metal for such a demonstration. The problem was that the experiment was complicated by the requirement that it had to be carried out using liquid helium and was therefore hard to set up and see. However, the discovery of high temperature superconductors with transition temperatures above that of liquid nitrogen (77K) means that now it is possible to show the effect in a very simple manner. The Meissner effect of complete magnetic flux exclusion is illustrated in Figure 2.5. This illustrates a small magnet floating continuously above a superconducting disc of yttrium Y – barium Ba – copper Cu – oxygen O, which is usually referred to as YBCO (this important ceramic material will be discussed in Chapters 3 and 4). Physics undergraduates at the University of Bath produced the sample. This high T_c material was kept in its superconducting state by cooling with liquid nitrogen held in an inverted polystyrene cup obtained from the University refectory!

Faraday's laws of electromagnetism tell us that when the magnet is moved towards the superconducting sample, a persistent or screening current is induced in the surface layer of the superconductor. This current in turn generates a magnetic field, which is exactly equal and opposite in sense to that of the magnet itself. Their exactly opposing magnetic fields cause the magnet and the superconductor to repel each other mutually with a force sufficient to result in levitation of the magnet, which stays floating above the superconductor in the spectacular manner shown in Figure 2.5. Since the superconducting current remains constant in the zero resistance material, the magnet will continue to float as long as the sample is kept below its superconducting transition temperature. Inside the bulk of the superconductor itself the magnetic flux is zero – as required by the Meissner effect.

Totalitarian Politics Intervenes

The 1930s saw dramatic political changes taking place in many countries as Europe lurched towards the outbreak of the Second World War. The increasing domination of dictators had an impact on almost every aspect of life, including that of the scientific community. As a result of the rise of the Nazi party in Germany in 1933, many prominent Jewish scientists fled that country; most of them emigrated either to the United States of America or England. Those coming to England from the low temperature physics community in Germany included the group from Breslau headed by Franz (later Sir Francis) Simon together with his nephew Kurt Mendelssohn as well as Nicolas Kurti and the brothers Fritz and Heinz London. All of these scientists went initially to the Clarendon Laboratory at Oxford having been invited by Frederick Lindemann, later Lord Cherwell, who was to become the chief scientific adviser to Winston Churchill and a powerful figure during the war years. The influx of these German scientists established the Clarendon Laboratory as a center of excellence for low temperature physics, which continues to this day. Mendelssohn was the first person to liquefy helium in the United Kingdom; this achievement enabled him to carry out a variety of investigations into superconductivity as well as some fundamental investigations into the nature of superfluid film flow in helium.

Bizarre circumstances and ruthless politics resulted in the beginning of low temperature physics in Moscow. In 1921, Lord Rutherford's highly prestigious Cavendish Laboratory in Cambridge was enhanced by the arrival of the brilliant and flamboyant Russian scientist Piotr Kapitza. Within a few years he became a professor of physics at Cambridge University and played a prominent role in the creation of the Royal Society Mond Laboratory devoted to investigations involving high magnetic fields and low temperatures. Kapitza produced liquid helium a short while after Mendelssohn had done so at Oxford. During more than a decade while he lived in England, Kapitza never took out British citizenship. Most years, he returned to holiday in Russia. In 1934, the Russian authorities refused to allow him to return to England after his vacation in the Crimea and Caucasus. A diplomatic incident occurred but Stalin was adamant that Kapitza must remain in Russia and carry out research in Moscow. After an initial period of singularly brave refusal, Kapitza finally was forced to stay and much of his equipment was transferred from Cambridge to Moscow.

Kapitza founded an outstanding Institute in Moscow devoted to the experimental aspects of low temperature physics. Life in Russia under Stalin was extremely harsh and dangerous: during the purges millions of innocent people were arrested, and most murdered, for alleged subversive activity. Not long after his return there, Kapitza with great courage demanded from Stalin that the brilliant young theoretical physicist Lev Davidovich Landau be released from prison where he had been placed for alleged anti-Soviet activity. Kapitza asked Landau to start up an Institute devoted to theory. This resulted in the formation of the Institute of Physical Problems, a small but outstanding group of theoretical physicists headed by Landau. Together, Kapitza and Landau spearheaded a Russian contribution to the development of both superconductivity and superfluidity, with a value that can hardly be overstated.

Separation into Type I and Type II Superconductors

During the second half of the 1930s it began to be appreciated that not all superconductors showed the comparatively simple type I behavior of *total flux exclusion* (Figure 2.4). Working at the newly created Royal Society Mond Laboratory in Cambridge, David Shoenberg examined the magnetic behavior of very small superconducting particles, notably colloidal mercury. His aim was to test the emigre London brothers' suggestion that although the magnetic flux would be excluded from the bulk of a superconductor, it would have to penetrate to a small depth in which screening currents flow. For sufficiently small particles, this penetration would occur in a significant fraction of the total volume and should result in the magnetic susceptibility becoming substantially less diamagnetic than for large samples. Shoenberg showed that this occurred in practice. Further evidence of departures from the ideal type I behavior was found in experiments with superconducting thin films carried out by Heinz London and E.T.S. Appleyard at the Mond laboratory and also at Bristol University; Alexander Shalnikov, working in Moscow at the Institute created by Kapitza, reached a similar conclusion. An exciting result was that much higher critical magnetic fields were found than would be expected for type I superconductors.

More striking evidence of departures from type I behavior was revealed by experiments carried out by Lev Shubnikov and his colleagues at Kharkov in Russia. They found that alloys like lead–bismuth required much greater magnetic fields to restore the normal state than those needed for the pure metallic elements. These alloys, and many others, which became known as type II superconductors, form a mixed state in which both normal and superconducting regions coexist (this is discussed in more detail in Chapter 8 and is shown in Figure 8.2). In this mixed state, a zero electrical resistance path through the superconducting regions still exists so that zero resistance is maintained. Type II superconductors can have substantially higher critical temperatures than type I. Furthermore, a magnetic field as large as several Teslas can be required to drive all of a type II superconducting sample normal. As a result, it has been possible to produce compact high field superconducting magnets using type II materials; this technology will be discussed in Chapter 8. Shubnikov was not to see the results of his pioneering work. He became a victim of the whims of the Stalinist regime. His arrest in 1937 led to the collapse of his research group and his own death in prison in 1945.

First Steps in Understanding Superconductivity

Largely as a result of the discovery of the Meissner effect, the 1930s saw the beginnings of an understanding of superconductivity. Initially it was appreciated that the occurrence of the Meissner effect demonstrated that superconductivity is a true equilibrium situation rather than a metastable one and represents a new phase quite distinct from that of a normal metal. The fact that the superconducting state is in equilibrium means that thermodynamic relations can be derived between the normal and superconducting states. The Dutch scientist Cornelius J. Gorter, who became a director of the Kamerlingh Onnes Laboratory in Leiden, pioneered the application of the principles of reversible thermodynamics to superconductors.

An important breakthrough occurred in 1934, when Gorter put forward the idea of a two-fluid model in which the electron gas within the superconductor is considered to have two components. One fluid component, which comprises the "superelectrons", is completely ordered and hence has zero entropy and carries the supercurrent, while the other component contains all the entropy of disorder and behaves like a normal electron gas. Below the superconducting transition temperature, the superconducting electrons short out the normal ones so that the electrical resistance is zero. A similar two fluid model has been proposed for super-fluidity in liquid helium, an example of the striking analogies that exist between the two phenomena, superconductivity and superfluidity.

Shortly after their arrival in Oxford in 1935, the brothers Fritz and Heinz London carried out a seminal theoretical analysis of the response of a super-conductor to an applied magnetic field. Their starting point was the newly established experimental finding that a bulk superconductor shows both zero electrical resistivity and perfect diamagnetism. To explain both of these observations simultaneously, they found it necessary to modify the classical Maxwell electrodynamic equations relating electric currents and magnetic fields. They derived new equations from which they were able to predict that although a magnetic flux is excluded from the interior of a superconductor, there is neverthe-less a shallow surface region having a characteristic distance, now known as the *penetration depth* λ, in which there is some penetration of flux. This magnetic flux

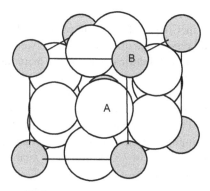

Figure 2.6 The A-15 (or beta-tungsten) crystal structure for compounds with the formula A_3B. Metals having this formula can be superconductors if A is a transition metal atom such as vanadium (V) or niobium (Nb) and B is usually an atom such as silicon (Si), germanium (Ge), aluminium (Al), gallium (Ga) or tin (Sn).

near the surface decays exponentially inwards. Shoenberg verified the ideas of the London brothers when he carried out his classic magnetic experiments on colloidal mercury, in which he was able to show that the temperature dependence of the number of superconducting electrons varied in a manner predicted by their theory. As mentioned earlier, these measurements reinforced the growing belief that type II superconductors should exist.

The Continuing Search for New Superconductors with Higher Critical Temperatures

From the early 1950s onwards, many new and technologically useful superconductors have been found. Pioneering studies, mainly in the large American industrial laboratories such as Bell Telephone, General Electric and Westinghouse, as well as in leading universities such as Stanford and Chicago, led to the discovery and later the production of new type II superconductors. Two leaders in this research were John Hulm and Bernd Matthias. John Hulm had obtained his doctorate from Cambridge University, where he had carried out research on superconductivity. In 1949, he went on to the Chicago Institute for the Study of Metals as a post-doctoral fellow. There he met Bernd Matthias, who was born in Germany, but studied in Switzerland, before gravitating to the United States. At the time that the two men met, Matthias was an assistant professor of physics at the University of Chicago on leave from Bell Telephone Laboratories. In the early 1950s, many people were investigating the physical properties of superconductors such as their specific heats, magnetic behavior and thermal and electrical conductivity. Taking a different approach, Hulm and Matthias studied superconductors from the viewpoint of chemists and metallurgists. They were especially interested in superconducting alloys and compounds containing the transition metal elements (which are characterized by possessing unfilled d shells in their electronic configuration; the nature of d states will be described in Box 13 in Chapter 11). Their work emphasised that, for super-conductivity to occur, the type of chemical bonding between the atoms, as well as the crystal structure and the metallurgical state of the material, is important. Their approach proved to be highly successful. Working either together, or in their separate research groups, they produced a large number of new superconductors. In particular they pioneered the A-15 structure (Figure 2.6) A_3B compounds, where A is often a $3d$ transition metal element and B is a non-transition metal element.

In 1954, Hardy and Hulm reported their discovery of the A-15 compound V_3Si, which has a transition temperature of 16.9K; the same year Matthias found Nb_3Sn with a transition temperature of 18.5K, a new record for T_c. For many years from the middle of the 1950s until the discovery of the first ceramic high temperature super-conductor in 1986, it was the intermetallic compounds with the A-15 structure that had the highest known values of the superconducting critical temperature T_c and critical fields. In consequence they have found widespread use in a small but flourishing superconductivity industry, mainly in the production of compact high field magnets operating up to several Teslas (see Chapter 8). Later it was found that as well as possessing high values of critical temperature T_c, these materials frequently show anomalies in physical properties such as their elastic constants; they undergo unusual softening in both their acoustic and optical vibrations, affecting their phonon population. This incipient lattice instability results in single crystals of Nb_3Sn and V_3Si undergoing a weak structural cubic–tetragonal transition at temperatures that are somewhat above T_c and has an influence on electron pairing (see Chapter 6) and hence on their superconducting behavior.

The approach to superconductivity adopted by Matthias merits further consider-ation: it throws interesting light on his research philosophy. He became increasingly disenchanted with theoreticians, many of whom he felt carried out over-sophisticated calculations and made extravagant predictions for the possible occurrence of superconductivity at high temperatures. He believed that such predictions bore no relation to the real world. By contrast, Matthias felt that his own approach never lost sight of reality. He had a prodigious knowledge and insight into the Periodic Table (Figure 2.1) and he combined this with a continual and almost obsessive hunt through the research literature to try and come up with materials that might turn out to be useful superconductors. As a result, he found a large number of new super-conductors. For many years his group at Bell Telephone Laboratories, and then at the La Jolla campus of the University of California, held the world record for the superconductor with the highest value of critical temperature. In his search for new and useful superconductors, he put forward several empirical rules, which indicated useful criteria for obtaining high temperature superconductors. His scorn for the outlandish predictions of some theoreticians is neatly summarized in a statement that he wrote in an article entitled "High Temperature Superconductivity?" for the journal *Comments in Solid State Physics* in 1970:

> *"In many discussions, I have tried to point out that the present theoretical*
> *attempts to raise the superconducting transition temperature are the opium*
> *in the real world of superconductivity where the highest T_c is, at present*

and at best, 21K. Unless we accept this fact and submit to a dose of reality, honest and not so honest speculations will persist until all that is left in this field will be these scientific opium addicts dreaming and reading one another's absurdities in a blue haze."

Matthias had a great respect and admiration for the seventeenth century astronomer and mathematician, Johannes Kepler, who derived the empirical laws for planetary motion, which Isaac Newton could explain using his theory of gravitation. In turn, Kepler had relied on the meticulous observations of planetary motion made earlier by Tycho Brahe. Matthias liked to think of himself as the Kepler of superconductivity, providing empirical laws and careful observations, which he hoped would allow a latter day Newton to construct a detailed microscopic theory having useful predictive powers for the occurrence of high temperature superconductivity.

Understanding Superconductivity

Despite the scorn for theoreticians shown by Bernd Matthias, the 1950s did see significant advances in the theoretical understanding of superconductivity, especially the successful formulation of a microscopic theory in 1957. In 1950, the first volume appeared of the classic book *Superfluids* by Fritz London that is devoted to superconductivity. In it he suggested that superconductivity is a quantum mechanical state observable on a macroscopic scale. In the same year, the Russian physicists Vitaly Ginzburg and Lev Landau extended the London brothers' phenomenological approach. This Russian work is based on Landau's theory of second order phase transitions, which introduces the idea of an order parameter. In the case of superconductivity, the order parameter distinguishes the superconducting phase from the normal phase. It is a complex wave function of the superconducting electrons. The Ginzburg–Landau theory provided new insights such as the dependence of the critical field and current on temperature and thickness in thin films. They introduced the concept of a *coherence length,* which gives a measure of the distance over which the superconducting wave function varies in zero magnetic field. The nature of the superconducting wave function itself was not clear. In 1953, using microwave techniques, which had been developed during the War years, Brian Pippard at the Mond Laboratory in Cambridge investigated the London penetration depth. He found that in order to reconcile his results with the London theory it was necessary to include the coherence length. Both the penetration depth and the coherence length are central to superconductivity and will be discussed further in this book. In 1957, the Russian Alexei Abrikosov was able to use the Ginzburg–Landau theory as the starting point for his own classic work, which predicted type II superconductivity and showed that in the mixed state the magnetic flux penetrates in a regular array of quantum vortices (Chapter 8). The era of the 1950s was the height of the "Cold War" between Russia and the West and there was little scientific interaction across the Iron Curtain. As a consequence many of these scientific achievements remained virtually unknown outside the Soviet Union.

Around this time experimental evidence was accumulating which suggested that an interaction between the electrons and atomic vibrations in a solid is inherent in the mechanism of superconductivity. A great leap forward in understanding was made in 1956, when Leon Cooper developed the important concept that superconductivity was associated with bound pairs of electrons travelling through the lattice. Each pair has equal but opposite spin and angular momentum. Building on this idea, John Bardeen and Cooper, together with a young postdoctoral assistant J. Robert Schrieffer, produced their famous BCS theory in 1957 in which superconductivity is considered to arise from the presence of these "Cooper pairs". Their comprehensive paper, published in the journal *Physical Review* in October 1957, is one of the most important and influential publications in condensed matter physics produced in the second half of the twentieth century. In an elegant manner, the BCS theory explained the known features of superconductivity, as well as making a number of predictions, which were subsequently shown to be true. Nevertheless, neither BCS nor any other theory has ever been able to predict which alloy or compound should have a high critical temperature T_c, as might have been expected for a truly comprehensive theory. Despite this, the BCS theory has provided us with a fascinating and clear insight into the cause and nature of superconductivity and is described in some detail in Chapter 6.

The quest for new high T_c materials has always been and remains to this day an empirical search with almost no theoretical guidance. In this respect, the disdain expressed by Matthias for theoreticians working in the field of superconductivity can be appreciated, although not condoned. However, there were also quite a few theoreticians with their feet firmly on the ground who were realistic about the chances of observing high temperature superconductivity. Prominent among these was John Bardeen, who had an excellent knowledge and grasp of the experimental situation in superconductivity. At one time he was interested in the possibility of excitonic superconductivity, whereby it might be possible to obtain pairing of electrons through electronic exchange rather than via phonons, resulting in high temperature superconductivity. Bardeen has written:

> *"In view of the large number of experiments that have been done and the wide variety of materials tested, many, including Bernd Matthias, have been pessimistic about the prospects for finding excitonic superconductivity. While these experiments do show that the conditions for observing it must be very exacting, they do not rule it out completely. Since the potential importance of high temperature superconductivity is so great, I feel that the search should be pursued vigorously even though the prospects for success may be small."*

Indeed this was in an article forming part of a *Festschrift* to celebrate the 60th birthday of Bernd Matthias.

A further fundamental advance took place in the early 1960s at Cambridge when Brian Josephson developed his far-reaching ideas on the tunnelling between two superconductors. What later became known as the Josephson effects have profoundly influenced our understanding of superconductivity. He investigated theoretically the behavior of two superconductors separated by a very thin insulating barrier, the sandwich now called a Josephson junction. Cooper pairs are able to tunnel through the barrier, which is then said to act as a "weak link". Josephson junctions have many applications, the most widespread of which involve the use of Superconducting Quantum Interference Devices: SQUIDS. The Josephson effects are discussed in Chapter 7 and some of their many their applications in Chapter 9.

Towards Higher Temperature Superconductors

In 1973, following on from the work of Matthias and Hulm on the A-15 compounds, John Gavaler produced a stoichiometric film of Nb_3Ge and observed a superconducting transition temperature of 23.1K, above the normal boiling point of liquid hydrogen; for a number of years this remained the record transition temperature. In the following years, all attempts to produce materials with higher transition temperatures were unsuccessful and this gave support to the widely held pessimism that superconductivity was confined to low temperatures and would be unlikely to occur much above 25K. The status of superconductivity by the middle of the 1980s was well summarized in an article on low temperature physics, which was part of a series produced by the US National Research Council under the title *Physics through the 1990s*:

> "The future of superconductivity in the next 10 years seems to be readily apparent from its past ten years. Even without the unpredictable discovery of a material with a much higher T_c, of an alternative pairing mechanism other than the electron–phonon interaction, or of new phenomena with the impact of the Josephson effects, the field is likely to continue to prosper along the lines of the recent past. It is reasonable to predict that new and unusual superconducting materials will continue to be discovered and avidly studied. Future improvements in the theory of dynamic phenomena in superconductors seem likely – so are improvements in device performance and in high field materials."

This article was first printed in April 1986, the same month that Georg Bednorz and Alex Müller, working at the IBM Zurich laboratories, submitted an innocuous looking paper to the journal "*Zeitschrift für Physik*". This paper was to transform completely our outlook on superconductivity and led to the most exciting period ever in condensed matter physics.

3 The Breakthrough to High Temperature Superconductivity

Scientists, engineers and industrialists were seized with excitement following the discoveries made in high temperature superconductivity in cuprates after the breakthrough in 1986 by Müller and Bednorz. Theirs was a most fruitful collaboration with consequences that must have far exceeded their wildest dreams. No well defined paths lead to great discoveries. Yet fortune favours the prepared mind. Their knowledge and expertise complemented each other and they made a shrewd decision to explore a class of materials not generally associated with superconductivity. The rest is history.

Choosing Perovskites

By the middle of the 1980s, Karl Alex Müller had already enjoyed a successful career in physics. Born in 1927, he had studied physics at the Swiss Federal Institute in Technology (ETH) in Zurich obtaining his doctorate in 1958. While there, he came into contact and was greatly impressed by the brilliant physicist Wolfgang Pauli who had won the 1945 Nobel Prize in physics for his discovery of the exclusion principle: fundamental to quantum mechanics. Müller saw Pauli as a role model. In 1963, Müller joined the research staff of the IBM Zurich Research Laboratory and by 1972 was head of physics there. Ten years later, he was promoted to an IBM Fellow. This prestigious appointment, awarded to few, gave him reign to work on any aspect of science that he wished. For almost his entire career, Müller had investigated a variety of problems in condensed matter physics. As far back as his doctoral studies, supervised by Georg Busch, he had investigated iron impurities in a recently synthesized oxide, strontium titanate ($SrTiO_3$), which has the ideal perovskite structure. His continuing interest in this oxide led him to focus on perovskite materials in general.

Strontium titanate is an example of an ideal perovskite ABX_3, which has the cubic crystal structure shown in Figure 3.1. Here the A and B atoms are metallic cations with a positive charge, while the X atoms are nonmetallic anions having a negative charge. The larger of the two metallic cations, denoted by A, lies at the

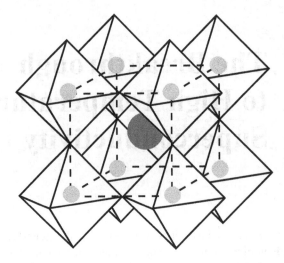

Figure 3.1   The basic perovskite ABX$_3$ structure. A is the larger metallic cation (shaded dark grey) at the center; B is the smaller metallic cation (shaded light grey) at the corners of the dotted cube. There is an oxygen atom (O is the atom X) on each corner of every octahedron; these are not shown.

center of a cube, and the smaller cations B occupy the corners. The anions X lie at the midpoints of the twelve edges. Over twenty elements have been found to occupy the A sites, around 50 elements the B sites. The anion X is frequently oxygen, although it may also be a member of the halogen family i.e. fluorine, chlorine or bromine. It can readily be appreciated that combinations of so many different possible ions can give rise to numerous ideal perovskite compounds. In addition, there are a large number of materials with a variety of modifications or defects, which depart from the ideal perovskite structure.

Perovskites were named in 1830 after the Russian mineralogist Count Lev Alekseevich von Perovski. They are naturally occurring minerals and, in the form of various silicates, are the most abundant materials in the earth's crust. From the point of view of condensed matter physics, they can be many things: insulators, semi-conductors, metals, superionics, ferroelectrics, piezoelectrics and others, and have been the starting point for much wide-ranging and fruitful research. The wide variety of physical properties can often be attributed to departures from the basic perovskite structure. For almost his entire career, perovskite materials have fascinated Müller. This obsession has led him to a faith in the ability of perovskite materials to generate new and important physics. It is a viewpoint that has served him well.

In 1978, Müller spent an eighteen-month sabbatical at the IBM Thomas J. Watson Research Center in Yorktown Heights, New York. There he turned his attention to superconductivity, a subject fresh to him. He was aware that there had been no major breakthrough in the area for several years: the highest known super-conducting critical temperature T_c had stubbornly continued to remain at around 23K. To exacerbate matters, IBM had recently abandoned its very costly and largely unsuccessful attempt to produce a marketable computer based on Josephson junctions (see Chapter 9), which are superconducting electronic devices capable of switching much faster than those based on semiconductors. Müller had read Michael Tinkham's classic book *Introduction to Superconductivity* but this did not provide him with any fresh approach that could clearly lead to new materials with higher critical temperatures. On his return to Zurich, Müller continued to investigate super-conductivity, at first working alone, and then, from 1983 onwards, with Johannes Georg Bednorz. The interests of the two men overlapped.

Bednorz was born in Neuenkirchen in Germany in 1950 and obtained his under-graduate degree from the University of Münster in 1976. He then worked for a PhD at the ETH in Zurich under the supervision of Müller and Professor Heine Granicher receiving his doctorate in 1982. As a result of his studies, Bednorz had developed wide-ranging expertise in areas such as crystallography, solid-state chemistry and condensed matter physics. One of his first experimental investigations was on the growth and characterization of single crystals of the perovskite, strontium titanate ($SrTiO_3$). This oxide is a well known ferroelectric. It was known that it can be made superconducting, if it is reduced so that oxygen atoms are removed from the lattice, but the critical temperature of around 0.3K is exceedingly low. Bednorz had already had some experience in superconductivity. In 1978, he spent a brief period at the IBM Zurich laboratory trying to enhance the superconducting behavior of $SrTiO_3$. Together Bednorz and Müller managed to raise its critical temperature to 0.7K by alloying with niobium.

For some while after they had teamed up, Bednorz and Müller worked in self imposed isolation. At that time, superconductivity was a rather unfashionable research topic with a poor track record of commercial success. Müller's position as an IBM fellow enabled him to pursue the topic without having to report to the managers at IBM. Both felt that if they were unsuccessful in their goal of finding a higher temperature superconductor, then the whole subject could be conveniently forgotten without jeopardizing Bednorz's career. At the outset, Bednorz and Müller realized that such an enormous amount of research had been carried out on inter-metallic compounds, especially the A-15 materials (see Chapter 2), that they were unlikely to make much progress in obtaining substantially higher critical tem-peratures by this route. Experience gained from their work on $SrTiO_3$ suggested that a more fruitful path might be to examine oxide materials.

A Lateral Move: the Discovery of Superconducting Oxides

The surprisingly high critical temperature T_c of 13K had already been discovered in 1973 in the perovskite-like lithium titanium oxide. X-ray analysis work showed this to be a complex multiphase system in which a spinel structure component was the one responsible for the occurrence of superconductivity. Of even greater significance was the discovery in 1975 of superconductivity, again at 13K, in a barium lead bismuth oxide $BaPb_{1-x}Bi_xO_3$ by a group at Du Pont, headed by Arthur Sleight. Since this material could readily be prepared in a single phase, as well as in the form of thin films, its potential for device applications attracted some interest.

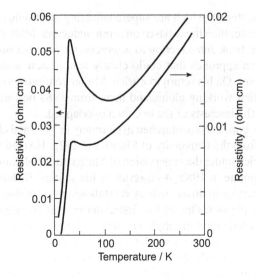

Figure 3.2 The temperature dependence of the electrical resistivity in two different La−Ba−Cu−O specimens. The steep drop towards zero was the first positive indication of high T_c superconductivity in cuprate materials. (Bednorz and Müller (1986).)

The Log-jam is Broken: Superconductivity Above 30K

In their search for oxide superconductors with even higher critical temperatures, Bednorz and Müller were guided by the BCS theory (see Chapter 6). For about two years they examined a large number of oxides, which were mainly nickel based, but failed to find any evidence for superconductivity. In the autumn of 1985, while Bednorz was carrying out a literature search, he became aware of a publication by Claude Michel, L. Er-Rakho and Bernard Raveau at the University of Caen in France. This paper described the behavior of some barium doped lanthanum cuprates, which have a perovskite structure, and exhibit metallic conductivity in the temperature range between +300°C and −100°C. The French group were chemists and their main interest was in the catalytic potential of the material. However, Bednorz and Müller felt that this system fulfilled many of the criteria which they believed were necessary for an oxide to show superconductivity, and so Bednorz set about preparing a series of solid solutions of the oxides. They began examining the samples in the middle of January 1986 and observed a metallic-like decrease in the resistance on initial cooling, followed by an increase (which in fact can be understood by a rather complex argument based on the effects of localization). Then a dramatic drop in the resistance by about 50% took place at 11K. The way in which the resistance varies with temperature follows the pattern shown in Figure 3.2. The sudden drop in resistance was the first tentative indication of a possible onset of superconductivity and led to a frantic effort on their part. By varying the sample composition, as well as the heat treatment, they were soon able to shift the onset of the resistivity drop up to 30K (Figure 3.2). Although this observation appeared to be highly significant, Bednorz and Müller were still by no means certain that what they had observed was genuine superconductivity. The history of this subject had been full of claims for superconducting transitions at high temperatures, which later turned out to be false. The dramatic drop in resistance could have been due to an entirely different physical effect, such as a transition from the metallic to an insulating state, or alternatively to something much more prosaic like the sudden appearance of an electrical short across their sample. Shades of Kamerlingh Onnes and his concerns so long before! They realized that they required evidence for the Meissner effect in their material before their claims would be taken seriously. To this end, they requested the purchase of a DC Squid Magnetometer for their laboratory in order to make magnetic susceptibility measurements. However, it was not until the middle of 1986 that this equipment arrived, was set up, tested and made ready to measure the susceptibility of the new materials. In the meantime, they carried out further extensive studies to try and identify the likely superconducting phase.

In April 1986, Bednorz and Müller submitted their paper to *Zeitschrift für Physik* entitled "Possible High T_c Superconductivity in the Ba−La−Cu−O System". This somewhat circumspect title reflected their recognition that they had not yet observed the Meissner effect; that is superconductivity had been indicated but not established conclusively. It was an elegantly written paper, which discussed the synthesis of metallic oxygen deficient compounds with the composition $Ba_xLa_{5-x}Cu_5O_{5(3-y)}$. They showed that samples with $x=1$ and $x=0.75$ and $y>0$, when annealed below 900°C under reducing conditions, produced three phases. One of these had a metallic, perovskite-type, layer-like structure that was related to the K_2NiF_4 structure; they suggested that it was this phase that might be responsible for any high temperature superconductivity.

This seminal paper appeared in the September edition of *Zeitschrift für Physik*. The required susceptibility measurements were carried out towards the end of 1986 in collaboration with a Japanese visitor, Masaaki Takashige. Some of the measurements are reproduced in Figure 3.3. The transition from a nearly temperature independent paramagnetic state to the strongly temperature dependent diamagnetic state with a negative susceptibility can be clearly seen. For Bednorz and Müller this observation of the Meissner effect confirmed that they were indeed observing superconductivity in these new ceramic materials.

The Immediate Impact of the Discovery

After their original paper appeared in *Zeitschrift für Physik*, Bednorz and Müller did not have to wait long before they realised that their work was being taken very seriously. The first rumours of outside interest began to circulate almost two months after publication, just before the start of the annual American Materials Research Symposium, which that year was being held in Boston, Massachusetts. One of those attending the meeting was the Japanese physicist Koichi Kitazawa from Tokyo University who had been invited to lecture on his latest work. However, for several

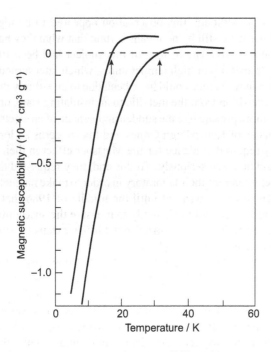

Figure 3.3 The low temperature magnetic susceptibility of La–Ba–Cu–O samples. The vertical arrows indicate the changeover temperature from paramagnetic to diamagnetic behavior; the marked diamagnetic behavior below that temperature indicates the onset of the Meissner effect, establishing that these materials are indeed superconducting. (Bednorz *et al.* (1987a).)

weeks prior to the meeting, Kitazawa, together with his group leader Shoji Tanaka and several colleagues, had abandoned their ongoing line of research and instead concentrated entirely on investigating the new superconductors. For some time prior to this, the Tokyo group had been examining superconductivity in the Ba–Pb–Bi–O system originally discovered by Arthur Sleight and his team at Du Pont. By the middle of 1986, the Japanese scientists had exhausted this line of research and were therefore most excited by the paper of Bednorz and Müller. In a series of communications to the Letters section of the *Japanese Journal of Applied Physics*, they confirmed and extended the findings of Bednorz and Müller. They demonstrated the Meissner effect as well as showing that the superconductivity occurred in the bulk material rather than being percolative in nature as originally suggested by Bednorz and Müller. They also identified the superconducting phase as having the composition $La_{2-x}Ba_xCuO_4$ with x in the region of 0.2. This material had a layered perovskite K_2NiF_4 structure that could be viewed as a stacking of $LaO–CuO_2–LaO$ sandwiches. They had the vision to suggest that the states responsible for the superconductivity probably lay in the CuO_2 plane at the centre of the sandwich and concluded that two-dimensional superconductivity is likely to be taking place, since the distance between neighboring CuO_2 planes is quite large. Later it was to be appreciated that the two-dimensional nature of the superconductivity is an essential feature of these new high temperature superconductors. Kitazawa presented some of the results from the Tokyo group on December 4, 1986, at the Boston meeting of the American Materials Research Symposium.

The Japanese were certainly not alone in showing avid interest in the new super-conductors. It quickly became apparent that groups in the United States, Europe, China and elsewhere had become actively involved. The possibility of competition from China was indicated when an article appeared on the front page of the Chinese *People's Daily* for December 27, 1986. It reported work by the team led by Zhong-Xian Zhao from the Institute of Physics in Beijing on superconductivity in the La–Ba–Cu-oxides. Perhaps of greater interest was the mention that the Beijing group had found evidence for a possible superconducting transition at around 70K, although tantalisingly no further details were given.

Since Bednorz and Müller worked at the IBM Laboratories in Zurich, a very active research program into the new ceramic superconductors was initiated at the two main IBM Research Laboratories in the United States. These are the Thomas J. Watson Laboratory at Yorktown Heights outside New York and the Almaden Laboratory near San Jose in California. Scientists in these laboratories rapidly began to carry out intensive investigations over a wide field of study, particularly with a view to potential commercial exploitation. They made some of the first studies on thin films of these materials.

Almost simultaneously, several research groups carried out experiments on materials in which barium was replaced by strontium, another alkaline earth metal. They all observed a further increase in the superconducting transition temperature up to 36K. Within a few days of each other, several papers were submitted for rapid publication. The Japanese group under Tanaka and the group at the AT&T Bell Laboratories led by Robert Cava and Bertram Batlogg were first. The findings of the people from the Bell Laboratories were printed on the front page of the *New York Times* for the December 31, 1986. This was symptomatic of a new trend in which many people heard about the latest breakthrough via the world's news media rather than through the scientific literature. This finding that the substitution of barium for strontium produced an increase in critical temperature T_c was also reported early on by the group at Bell Communications Research Laboratory (Bellcore) and in Zurich by Bednorz, Müller and Takashiga; the results of these studies appeared in print during February 1987.

Superconductivity Above the Boiling Point of Liquid Nitrogen

A major player in high temperature superconductivity has been Paul (C W) Chu who, during these exciting times, was working at the University of Houston in Texas. Chu has had a long and distinguished career of research into superconductivity; he is particularly well known for his investigations into the effects of pressure on the superconducting transition temperature T_c. He was born in the Hunan province of China in 1941 and named Ching-Wu. His parents were members of the Nationalist Party and they became involved in the civil war in China, which began in 1945. Life was very difficult for the family and they were forced to leave China and flee to Taiwan in 1949.

Chu's scientific education began in Taiwan and in 1962 he graduated from Chengkung University. Like many ambitious students in Taiwan at that time, he wanted to pursue his graduate studies in the United States and managed to enrol for a Masters degree at Fordham University in the borough of Bronx in New York. On arrival in America, Chu began to adopt his Christian name Paul. He studied under Joseph Budnick, a distinguished solid-state physicist. His career began to advance after he had embarked on his doctoral studies at San Diego in California under Bernd Matthias. Chu and Matthias were kindred spirits. Like Matthias, Chu is an outstanding scientist who is able to think creatively and imaginatively and has the self-confidence to follow through his ideas. Trying to find materials with ever-higher superconducting transition temperatures obsessed both men. So far, success had largely eluded Chu. Some years earlier he was involved in a controversy over possible superconductivity in copper chloride (CuCl). While working at Cleveland State University in Ohio, Chu collaborated with a Russian visitor Alexander Rusikov. Together, they studied CuCl and repeatedly observed anomalies in the magnetic behavior around 200K, which they speculated might be attributed to the onset of superconductivity. On returning to Russia, Rusikov, somewhat unwisely, announced that they had observed superconductivity at this temperature. This was widely reported. Subsequent measurements by scientists at the Bell Laboratories showed conclusively that pure CuCl was not superconducting, although Chu continued to argue that superconductivity might be occurring in the impure material. However, the whole episode probably did little for his scientific reputation.

It was on returning to Houston early in November 1986, having spent the previous month at the headquarters of the National Science Foundation in Washington D.C., that Chu first saw a copy of the paper by Bednorz and Müller, which one of his students had left out for him to read. For Chu, this was to be a defining moment in his life. He realised at once that the new La–Ba–Cu–O ceramics should be examined under high pressure. Chu had developed a highly effective research group and he and his team launched themselves into a concerted effort to investigate the new materials. They quickly demonstrated that application of pressure produced a dramatic increase of the transition temperature. No other material that they had ever previously studied had shown such a large increase in the super-conducting transition temperature with increasing pressure. Chu presented some of his group's first results at the Boston meeting of the Materials Research Symposium on December 4th. Some of the data from their publication in *Physical Review Letters*, which appeared early in 1987, are shown in Figure 3.4, where it can be seen that the superconducting transition temperature was increased to around 40K by an applied pressure of 13 kbar. Further application of pressure led to a rise in the transition temperature up to 52K. Quite apart from setting new records for T_c, their work had enormous significance: it pointed clearly to the direction in which to proceed. What was needed was to find a completely new compound made from smaller atoms that could simulate at atmospheric pressure the "squeezing" effect due to the application of pressure. This clever idea directed a frenzied effort by Chu and his colleagues to find such a material.

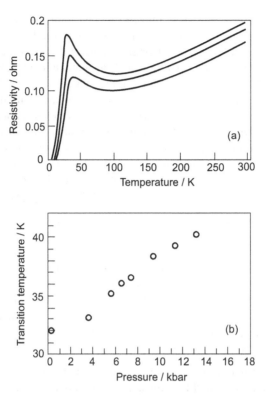

Figure 3.4 (a) Temperature dependence of the electrical resistivity at selected pressures for a La−Ba−Cu−O sample. Upper curve: 0 kbar, middle curve: 6.6 kbar, lower curve: 13.3 kbar. (b) The transition temperature T_{co} at which superconductivity onsets as a function of pressure for the same La−Ba−Cu−O sample. (Chu *et al.* (1987).)

Chu realized at the start of his research into high temperature superconductivity in the oxide ceramics that there was going to be fierce competition between several leading groups but enormous rewards for those who were first successful. He appreciated that to achieve success it would be advantageous to have additional skilled experimentalists working on the problem. While he was at the Boston conference, Chu met up again with one of his former students, Maw-Kuen Wu. Chu invited him, along with his research group, to join him in the search for the new superconductors. Like Chu, Wu had come to the United States from Taiwan and was also prepared to strike out boldly in search of new superconducting materials. He led a research group, which specialized in preparing and studying oxide materials, at the Huntsville campus of the University of Alabama. A concerted effort on the part of both groups rapidly led to success. The decisive move was the decision to replace lanthanum with the element yttrium. Although this had been mentioned as a possibility in discussions between the two groups, it appears that the main impetus to do so stemmed from some empirical calculations made by Jim Ashburn, who was one of the graduate students of Maw-Kuen Wu. Together with another graduate student Chuan-Jue Torng, they made an yttrium (Y) – barium (Ba) – copper (Cu) – oxide sample. At their first attempt made on January 29th 1987, they saw evidence for superconductivity at 93K in their material. In great excitement, they telephoned Chu and the following day, Wu and Ashburn flew to Houston with their sample to carry out further investigations. Chu contacted the editor of *Physical Review Letters* and immediately wrote and sent off two back-to-back manuscripts to that journal, which arrived at the offices of the American Physical Society on February 6th. Ten days later, Chu held a press conference announcing the discovery but withholding any details about it, and instead, referred all inquiries to the forthcoming publications. Meanwhile, he and his colleagues had filed for a patent. The papers appeared in the March 2nd edition of *Physical Review Letters*. The two groups together reported superconductivity up to 93K, well above the boiling point of liquid nitrogen (77K), in a ceramic material comprised of Y–Ba–Cu–O. The opening lines of their first paper are emphatic:

> "*The search for high-temperature superconductivity and novel super-conducting mechanisms is one of the most challenging tasks of condensed-matter physicists and material scientists. To obtain a superconducting state reaching beyond the technological and psychological temperature barrier of 77K, the liquid-nitrogen boiling point, will be one of the greatest triumphs of scientific endeavor of this kind... ... ... For the first time, a 'zero-resistance' state ($p < 3 \times 10^{-8} \, \Omega \, cm$, an upper limit determined by the sensitivity of the apparatus) is achieved and maintained at ambient pressure in a simple liquid-nitrogen Dewar.*"

It can be seen that, in stark contrast to the tentative wording of the paper by Bednorz and Müller, Chu and his colleagues were anxious to dispel any lingering doubts that might be held by others that they had indeed observed superconductivity at the temperature that they claimed. They took pains to emphasize that they were the first people to observe zero resistance in a material immersed in a simple liquid nitrogen Dewar. The superconducting transition temperature that they observed decreased in an applied magnetic field as would be expected for genuine super-conductivity. Their results for the temperature and magnetic field dependence of the resistance are reproduced in Figure 3.5. Also shown are their magnetic susceptibility measurements, which clearly show the Meissner effect. Very significantly, the first reference in the first paper was to a United States patent application for their discovery. Chu had made strenuous efforts to ensure that the details of his discovery were kept secret until these papers had been published. Nevertheless, information about the discovery was leaked, although the perpetrator was never traced. However, Chu was fortunate in that in his great haste to write and submit the manuscript, he had inserted the symbol Yb for ytterbium instead of Y for yttrium. This error was only corrected during the proofreading stage. Chu has always maintained that this was a genuine error resulting from the manuscript having been produced so rapidly. Others, who were misled by the leaked information, have accused him of having deliberately inserted the wrong symbol. For Chu and others much was at stake. Without doubt, there was the heady possibility of being awarded a Nobel Prize. More importantly, in the long run, were the enormous technological, and hence com-mercial, implications since it was clear that a superconducting technology based on liquid nitrogen would be so much cheaper and easier to use than that based on liquid

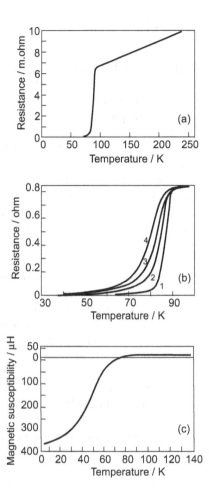

Figure 3.5 The first observation of a transition to the superconducting state above the boiling point (77K) of liquid nitrogen as shown in (a) the temperature dependence of the electrical resistance, (b) the effect of application of an increasing magnetic field (1–4) on the resistance and (c) the temperature dependence of the magnetic susceptibility (demonstrating the Meissner effect) of a Y–Ba–Cu–O sample. (Wu *et al.* (1987).)

helium. The prospect of lucrative markets lay ahead. It was essential for Chu and his colleagues to try and establish priority for their discovery. They were certainly none too soon. On February 25, 1987, the Chinese *People's Daily* announced the discovery by the Beijing group of superconductivity at 90K in Y–Ba–Cu–O. This was widely reported around the world. In addition, several groups in Japan acted on this announcement and within a few days, managed to reproduce the results. The group in Tokyo even claimed that they had first observed superconductivity around 90K as early as February 20th. The discovery of the 90K superconductor, coming so soon after the initial discovery of Bednorz and Müller, had certainly stimulated further research.

The New Era is Ushered In

The weeks and months immediately following the discovery of high temperature superconductivity were exciting and unprecedented ones for condensed matter physicists throughout the world; many vied with each other to reproduce and extend the work. As articles were hurriedly prepared and rushed into print, many scientists temporarily discarded the methodical, scientific approach in which research articles are carefully written and peer reviewed. Several new journals spawned with the sole aim of the rapid dissemination of new results. Faxes, preprints, telephone conversations and press releases became the norm as people and companies competed with each other to present new results and establish priorities for their discoveries. There has never been anything quite like it in the history of science.

An early indicator of the atmosphere at that time was the famous Extraordinary Session of the March 1987 meeting of the American Physical Society. This was held in New York and the session took place on the March 18, barely two weeks after the appearance of the Chu paper. This was the first opportunity for many scientists to report their new findings. Over two thousand scientists attended the session and an additional two thousand saw it relayed on video in one of the anterooms. The chairman was Brian Maple from La Jolla, California and the format was that five people from leading research institutes in the field were each allocated twelve minutes to talk and answer questions followed by anybody else having a slot of five minutes.

The five principal speakers were Alex Müller from IBM in Zurich, Shoji Tanaka from the University of Tokyo, Paul Chu from Houston, Zhong-Xian Zhao from the Institute of Physics in Beijing and Bertram Batlogg from AT&T Bell laboratories. This was the first opportunity for most people to be able to listen directly to what Chu and his colleagues had achieved, and understandably, there was enormous interest in his presentation. However, in many respects, the star of the show was Bertram Batlogg who dazzled the audience with a series of stunning viewgraphs. Towards the end of his talk, he showed a one inch diameter toroid made from the new superconducting material, which could readily form the core of a superconducting motor or magnet. Immediately afterwards, he produced a foil of the superconductor suitable for manufacturing a variety of electronic devices. The progress that had been achieved in such a short space of time by the group from AT&T Bell laboratories was awesome. Batlogg concluded his presentation with the emotive words "*I think our life has changed*", whereupon the audience erupted with loud cheers and clapping.

The five main speakers were followed by a host of others all anxious to present their results and hopefully establish some sort of precedence for their discoveries. The main areas of interest were: the composition and structure of the new super-conductors, especially the 93K materials; how far these materials could be reconciled with the BCS theory; possible new superconductors with even higher transition temperatures and the whole range of technological applications. Speaker followed speaker until the meeting finally drew to a close in the small hours of the morning. Few were in doubt that the evening session had represented a landmark in the history of science and possibly even marked a turning point in human development. The event subsequently became known as the *"Woodstock of Physics"*. It was widely believed that the recent discoveries ushered in far reaching new physics and a revolutionary era in which superconductors would soon underpin everyday technology.

Rapid Award of the Nobel Prize

After a brief period of scepticism, it was universally recognized that Bednorz and Müller had made a major discovery. Its impact was such that they received the Nobel Prize for physics the following year, 1987. This was one of the most rapid awards ever given for this prestigious prize and was one of the rare occasions that the original stipulation of Alfred Nobel's will was fulfilled exactly in that the award should be made to those who "during the past year, shall have conferred the greatest benefit on mankind". The Nobel Prize citation noted that their discovery inspired a great number of scientists to work with related materials. This would seem to be a considerable understatement. During 1987, a substantial proportion of condensed matter physicists began working in this field. There were many people who felt that Paul Chu also deserved to be included in the Nobel Prize award. It has been argued that Bednorz and Müller had a head start of several months over Chu and his colleagues but failed to find a superconducting ceramic above the boiling point of liquid nitrogen. Others have argued that Chu's breakthrough is of technological rather than scientific significance. Chu also became embroiled in controversy with the researchers from Alabama over priority for the discovery of the 90K superconductor. There is little doubt that the first people to observe superconductivity above 90K were Maw-Kuen Wu, Jim Ashburn and their colleagues at the University of Alabama. This was specifically recognized in the first of the two seminal publications in *Physical Review Letters* that appeared on March 2, 1987, in which Wu was the first author. It is equally true that it is highly unlikely that they would have made the discovery without the input from Chu's group at Houston. It had been their pressure measurements on La$-$Ba$-$Cu$-$O that clearly pointed in the direction for future investigations. It was natural that Chu, a very senior scientist who spoke excellent English, should act as spokesman for the two groups. Since Wu had been a former student of Chu's, it was understandable that he would tend to defer to his former research supervisor. However, just as Gilles Holst may merit much greater credit for the initial discovery of superconductivity in 1911, so Wu and Ashburn could well deserve greater recognition for the discovery of Y$-$Ba$-$Cu$-$O. In addition, it is unclear how much credit should go to the Tokyo and perhaps the Beijing groups for the discovery of the 90K superconductor. The Nobel Prize committee astutely avoided any controversy by awarding the prize just to Bednorz and Müller. Since the deadline for nominations is February 1st, the prize could easily be awarded solely to them since it was a month before the publications of Chu and colleagues.

There are some interesting parallels between the initial discovery of superconductivity in 1911 and that of the ceramic superconductors by Bednorz and Müller 75 years later. In both cases, the theoretical arguments, which gave rise to each investigation, were questionable. Kamerlingh Onnes was testing an idea put forward by Lord Kelvin in which the latter suggested that at extremely low temperatures the electrons were likely to attach themselves to the positive ions resulting in an increase in the resistance and eventually the conductor becoming an insulator. As a result of investigating this idea, Kamerlingh Onnes discovered superconductivity, an outcome which nobody could have possibly predicted from the existing understanding of metals at that time, but which proved to be much more exciting and far reaching than his original experimental objective. In fact, there has been no evidence at all for Kelvin's suggestion. Although Bednorz and Müller set out in a deliberate search for high temperature superconductors, their speculations as to the criteria necessary for possible candidates were largely insubstantial. They placed enormous faith in materials that had the perovskite structure. They also worked within the framework of the BCS theory and attempted to identify metallic oxides in which the electron–phonon interaction was likely to be enhanced. However, we now know that the two dimensional nature of the cuprate materials is a much more important condition for the occurrence of high temperature superconductivity than their original ideas. Nevertheless, Bednorz and Müller made a discovery of enormous significance. In both the original finding of superconductivity and then that of the high temperature superconductors, there is no doubt that serendipity played an important role. The history of science is full of this. The discovery of x-rays and radioactivity are two important examples from the end of the 19th century while nuclear fission and the cosmic background radiation are from the 20th century. Many books on science tend not to emphasize enough how often scientists make important discoveries through a chance observation. With cutbacks in funding for research in both universities and industry likely to continue, less and less money is made available for speculative "blue sky" research. In the long run, this is detrimental to innovative thinking. The policy adopted by IBM of giving scientists, with an outstanding track record of successful research, the status of IBM Fellow, enabling them to pursue their own lines of research largely unimpeded by constraints, proved singularly successful. In this respect, it is significant that earlier the 1986 Nobel Prize in physics had been awarded jointly to Gerd Binnig and Heinrich Rohrer, also of the IBM Laboratories in Zurich, for their discovery of the scanning tunnelling microscope.

Onwards and Upwards

The discovery of materials with transition temperatures well above that of liquid nitrogen finally laid to rest the widely held belief that superconductivity is confined to the temperature region close to the absolute zero. It spurred efforts to find materials with even higher transition temperatures – possibly even as high as room temperature. A good indication of the dramatic increase in the value of the critical temperature T_c in the few months following the original discovery of Bednorz and Müller can be seen from Figure 3.6. In the 75 years since the initial discovery of superconductivity by Kamerlingh Onnes in 1911, the value of T_c rose by less than

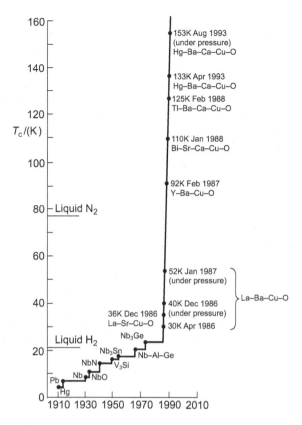

Figure 3.6 How the maximum superconducting transition temperature T_c has changed with time, showing the dramatic increase resulting from the international scientific response to the discovery of the high T_c materials. (Bednorz and Müller (1988).)

20K, whereas in under a year following the work by Bednorz and Müller, it rose by over 70K with a further increase of about 30K early in 1988.

A year after the important discovery of superconductivity above the normal boiling point of liquid nitrogen (77K), interest became focused on two similar series of compounds both of which showed superconductivity above 100K. The French group at Caen led by Bernard Raveau were the first people to show the existence of superconductivity in the system Bismuth (Bi) – Strontium (Sr) – Copper (Cu) – Oxygen (O) at a temperature of around 10K. In January 1988, a Japanese group working at the National Institute for Metals in Tsukaba led by Hiroshi Maeda first reported the existence of a superconducting phase with a T_c of around 110K in Bi–Sr–Ca–Cu–O, that is where Calcium Ca had been added to the system studied by the French. The structure and properties of this system were intensively studied by several groups; it was quickly established that there was a new series of super-conductors with the structural formula $Bi_2Sr_2Ca_{n-1}Cu_nO_{2n+4}$ with T_c of 10K, 85K and 110K for $n = 1, 2$ and 3 respectively (see Chapter 4). One reason for the interest in this system was that it represented the first high temperature superconductor that did not have a rare earth element in it. It also focused attention on the importance of the two dimensional layer-like structure. The bismuth superconductors have been studied extensively and have features that make them useful for technological applications.

Less than a month after the discovery by the Japanese group of the bismuth based superconductor, Allen Hermann and Zhengzhi Sheng of the University of Arkansas in the USA announced superconductivity above 100K in the thallium (Tl) based system Tl–Ba–Ca–Cu oxide. Again, several research groups established that the structural formula for this system is $Tl_2Ba_2Ca_{n-1}Cu_nO_{2n+4}$ with T_c of 80K, 110K and 125K for $n=1$, 2 and 3 respectively. Less is known about the thallium than the bismuth compounds, in part because of the extremely toxic nature of thallium. A critical temperature T_c of 125K for the $n=3$ thallium compounds was the highest superconducting transition temperature for a few years despite enormous efforts to increase the transition temperature and several claims for higher temperatures which later have turned out to be false. However, in 1993, the group at the Technical Hochschule in Zurich, led by Hans Ott, reported superconductivity at above 130K in the system Hg–Ba–Ca–Cu oxide. This is yet another layered material with CuO_2 planes in which the thallium is replaced by mercury (Hg). Further work by Chu and his colleagues, in which the compound $HgBa_2Ca_2Cu_3O_{8+\delta}$ was subjected to pressures up to 150 kbar, led to a T_c in excess of 150K. At present (2004) this would seem to be the current record for the superconducting critical temperature.

4 Structure of the Cuprates. CuO₂ Layers: the Genie for High Temperature Superconductivity

The Defective Nature of Cuprates

Once superconductivity had been clearly established in the new ceramic oxide materials, such as YBCO, the race was on to determine their composition, structure and basic properties. Few materials have ever been subjected so intensively to experimental study, using every conceivable technique. Scientists and engineers have a habit of saying "Nothing is easy". This is particularly true of work on the cuprates. It has proved singularly difficult to make good clean material. After fabrication, samples always contain many defects, including grain and twin boundaries, vacancies and dislocations. The atomic arrangements are by no means perfect; the layers are not always stacked in the ideal sequence. In addition the constituent elements are often not in the exact atomic proportions suggested by ideal chemical formulae; a deficiency in oxygen is not unusual, and this can completely change the electronic properties, including superconducting behavior.

As a result of the highly defective nature of the materials, inconsistencies between experimental results abound, making it difficult to get a clear picture of what is going on. Such a situation is not rare in science, especially during the period immediately following a major discovery that generates a huge amount of research, as it has for the cuprate superconductors. Over forty years ago, in the concluding section of a comprehensive review article on "The ordinary transport properties of the noble metals", the eminent theoretician John Ziman wrote:

> *"It is not easy to sum up this complicated subject. The noble metals have suffered so many experimental observations, which have been subjected to so many theoretical explanations, that one can always find some evidence to support any point of view. Many of the observations are inconsistent with one another and with all theories. The abnormal sensitivity of many properties to small amounts of impurity leads to conflict of evidence, which only further deliberate experiments can resolve. It is not merely a question*

*of fitting together a jig-saw puzzle; every piece has several spurious
versions which must be identified and discarded."*

Ziman's words echo down the years and are particularly apt in the case of the high
temperature superconductors.

For many years superconductivity belonged almost entirely to the domain of the
physicist. By contrast, a striking feature of the research into high temperature superconductivity has been the wide range of backgrounds of the people at work on it:
neatly summarized by Ted Geballe and John Hulm, two leading physicists in the
development of type II superconductors, in an early review article:

> *"The nature of the research, which requires physicists, materials scientists,
> electrical engineers, chemists and ceramicists, is a textbook example of the
> vitality of interdisciplinary research – the hallmark of modern materials
> research".*

A typical example of somebody who was drawn into studying high temperature
superconductors is Robert Hazen. He works at the Geophysical Laboratory of the
Carnegie Institution at Washington, which concentrates on the study of the physics
and chemistry of earth materials: minerals, rocks, volcanoes and the earth's deep
interior. Hazen is an expert crystallographer and was asked to study the structure of
one of the first samples of YBCO produced by Paul Chu. The opportunity transformed his life, and even that of his family for a time, as he recounts in his book
Superconductors: The Breakthrough.

Structure and Doping of $La_{2-x}Ba_xCuO_4$

Whenever a new material is discovered, one of the first things to do is to find out how
the atoms are arranged in it. Such structural work is usually carried out using x-ray
and neutron scattering techniques. Although the first cuprate superconductor
$La_{2-x}Ba_xCuO_4$, found by Bednorz and Müller, had a known crystal structure, this
was not true of the other layer-like cuprate compounds such as YBCO, which were
previously unknown and had atoms in arrangements that had not been observed
before. Knowledge of the structure has turned out to be crucial to understanding the
mechanism of superconductivity in the cuprates. So as each new high temperature
superconducting compound has been discovered, its structure has been subjected to
intense scrutiny by crystallographers. From a practical point of view, realization that
the cuprate superconductors $La_{2-x}Ba_xCuO_4$ and YBCO have layer-like structures
was a starting point for the search for related compounds made from other elements;
work stimulated by that approach led to the development of series of bismuth,
thallium and mercury compounds, which showed progressively higher critical temperatures climbing up to the present record (Chapter 3).

The most striking feature, common to all the high temperature cuprate superconductors, is a layered perovskite-like structure whose building unit contains one or
more copper oxygen layer planes often just referred to by its chemical formula CuO_2.

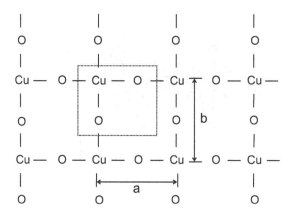

Figure 4.1 A single CuO_2 plane. Each copper (Cu) atom is bound to four oxygen (O) atoms in a square planar configuration. In the layer-like structure of cuprates superconductivity and carrier transport are mostly confined to these CuO_2 planes. The box with the dotted outline defines the repeating CuO_2 unit. The cell dimensions are a and b. In a three dimensional tetragonal crystal a and b are equal (see Figure 4.4) but in an orthorhombic material they are not. The CuO_2 layers lie in the ab-plane of the structure.

A flat CuO_2 plane is shown in Figure 4.1. In some compounds, these CuO_2 layers are puckered. Electrical conduction and superconductivity are largely confined to these layers, that are parallel to the ab plane. By comparison conduction in the direction, denoted as the c-axis, at right angles to the layer planes is small; therefore, so far as the conduction of electricity is concerned, the materials can be thought of as being quasi-two-dimensional.

The initial discovery of high temperature superconductivity made by Bednorz and Müller was in barium doped lanthanum copper oxide La_2CuO_4. The crystal structure of the pure, undoped material is shown in Figure 4.2. It is related to that of a class of materials with the general formula K_2NiF_4. Materials having such structures, made from a variety of atoms, have been quite actively studied for a long time and found to exhibit a wide range of electrical and magnetic properties. In the crucially important CuO_2 layers in La_2CuO_4 each copper atom is strongly bonded to four nearest neighbor oxygen atoms (labelled O_1) in a square. These squares are linked together to form the CuO_2 layers; the copper–oxygen distance of about 1.90Å is rather short, indicating strong covalent bonding between the atoms. The CuO_2 planes have two La–O planes between them and so are spaced quite far apart at a distance of 6.6Å. This separation severely limits information exchange between one CuO_2 layer and its neighbor, causing the material to have a two dimensional character.

A crucial property of the cuprates is that they can be doped; that is, some of the atoms in the pure form are replaced by different ones. Doping has dramatic effects on the electrical properties of $La_{2-x}Ba_xCuO_4$. The undoped (pure) compound La_2CuO_4 is an insulator; the material only shows metallic behavior as a result of the addition of excess charge. This is achieved by substitution of trivalent La^{3+} cations by divalent Ba^{2+} cations, which supply only two electrons, compared with the three

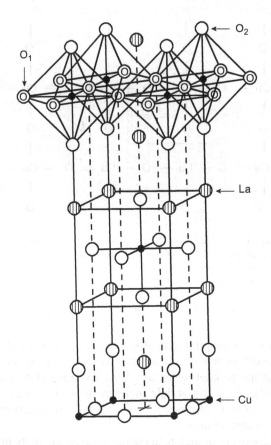

Figure 4.2   The structure of pure La_2CuO_4. Small filled circles: Cu; open circles: O; circles with vertical hatched shading: La. To show a CuO_2 layer which contains the superconducting electrons, a smaller circle has been placed at the centres of the appropriate oxygen atoms (these particular atoms are often labelled O_1). Each copper atom is actually at the centre of an octahedron of oxygen atoms, as shown for a few cases at the top. The copper atom is bound more weakly to the two more distant, apical oxygen atoms (often labelled O_2) that are on the top and bottom of the octahedron. These oxygen atoms labelled O_2 are associated with La–O planes, which are inserted between (or intercalated with) the CuO_2 layers.

donated by the lanthanum ion La^{3+}. In the jargon of semiconductor physics, the dopant atoms act as acceptors and introduce additional "holes" into the conduction band associated with the CuO_2 planes, thereby giving rise to conducting behavior and the possibility of superconductivity. To make a superconductor requires a substantial number of free carriers. Therefore many dopant atoms are needed to induce a metallic-like conducting state. For $La_{2-x}Ba_xCuO_4$, superconductivity does not occur until the barium dopant concentration reaches, or exceeds, a value x of 0.06.

In many metals, electrons are the carriers of electricity but the truth of the matter is that in most cuprates the carriers of electric current are the "holes", which can be thought of as positively charged spaces left behind when electrons are removed. Conduction can be thought of as taking place by electrons hopping from one hole to another. Holes are formed when dopant atoms, such as Ba^{2+}, have a lower valency than the atoms that they replace, like La^{3+}: the Ba^{2+} ions added to the materials soak up one electron each. It is common practice to use the term "carriers" rather than "electrons" when describing the entities, which make up the Cooper pairs in many cuprates. A crucial step came with the finding that there are cuprates, such as $Nd_{1.85}Ce_{0.15}CuO_{4-y}$ in which the carriers are electrons in the usual sense, which nevertheless are superconductors. This shows that the high temperature superconductivity in most of the other cuprates does not simply arise from hole-like behavior of the conducting electrons as many once thought. Nevertheless, there are far more high temperature superconducting cuprates, which have "holes" rather than "electrons" as carriers.

Preparation of YBCO

The frenetic search for superconducting cuprates with a high transition temperatures quickly led to the discovery of YBCO with a transition temperature of around 93K by Chu and colleagues at Houston (see Chapter 3). Considerable attention has been focused on YBCO because it is the first superconductor to be discovered which can be used in liquid nitrogen (boiling point 77K). The economic advantage of this can be seen by noting that the difference in price between liquid nitrogen and liquid helium is similar to that between a modest beer and a good malt whisky. On top of this, a helium system is inherently complex and prone to breaking down: a technology based on liquid nitrogen as a refrigerant is enormously attractive and is now in place.

YBCO remains the best known and most widely used of the high temperature superconductors. One reason for its popularity is that it is possible to make a polycrystalline sample of this material by a reasonably straightforward and non-hazardous process: even carried out in high schools. It is a unique experience that recent research, based on Nobel Prize winning work, can be replicated by school children! It does much for the popularization of science that the levitation produced by a superconductor on a small magnet, as shown in Figure 2.5, can so easily be demonstrated to a general audience such as children or members of the Women's Institute. Personal experience has shown that both these groups of people make appreciative, lively and to say nothing of questioning audiences at lecture demonstrations that include high temperature superconductivity.

To make YBCO (i.e. $YBa_2Cu_3O_{7-x}$), the procedure usually followed is first to mix together the right proportions by weight of the oxides Y_2O_3, BaO and CuO to achieve the required atomic ratio $Y_1:Ba_2:Cu_3$: one atom of yttrium to two of barium to three of copper: i.e. 123. In practice, because the oxides of the alkaline earth metals such as barium can be unstable in air, the more stable carbonates are frequently used instead as starting materials. Mixing is normally carried out using a ball-mill but a pestle and mortar will do. The mixture of components is reacted at a temperature of between 800 to 850°C for a period of ten hours. This process (technically called calcining) may be repeated several times so as to obtain the

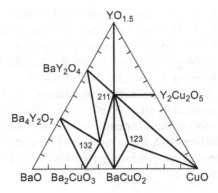

Figure 4.3   The phases that can be made starting from barium oxide BaO, copper oxide CuO and yttrium oxide Y_2O_3 are shown on this pseudoternary phase diagram, which has been prepared at a temperature of 950°C in air at which the phases are stable. To keep the numbers of cations consistent, Y is needed rather than Y_2; therefore $YO_{1.5}$ is shown rather than Y_2O_3 on one of the three apices. The important high temperature superconductor $YBa_2Cu_3O_{7-x}$ is labelled as 123. The green phase is shown as 211. The phase diagram shown here is taken from the review article by Beyers and Ahn from the I.B.M. laboratories at San Jose and was compiled using several sources.

optimum cation stoichiometry (that is the required atom ratio 123) and improve the superconducting properties. For demonstrations of the Meissner effect, the superconductor is best made in the form of a disc. To make such a pellet, the YBCO powder is put into a die and compressed, using a simple hydraulic press, to a pressure of a few kilobars. The pellets are then given a further heat treatment at a temperature of about 850°C for a day to homogenize them. The superconducting behavior of $YBa_2Cu_3O_{7-x}$ is extremely sensitive to the oxygen content (this will be discussed later and is shown in Figure 4.6): a requirement for a high critical temperature is that the oxygen content is close to O_7, which is the fully oxygenated state with x equal to zero. So the pellets are given an "oxygen soak", that is they are maintained at a temperature of around 430°C for twenty-four hours in air or oxygen, a treatment that results in a material of composition close to the desired O_7 value.

Metallurgists and materials scientists usually represent the range of temperature and composition over which compounds or alloys are stable on a phase diagram. Construction of such a diagram is the usual starting point for scientists making and studying new materials, especially in systems complicated by a number of different phases. This is a useful step here because several stable compounds can be made beginning from BaO, CuO and Y_2O_3. To construct a ternary phase diagram (Figure 4.3), these starting oxides are placed at the corners of an equilateral triangle. Compositions of mixtures of two of these oxides are then plotted along each edge of the triangle. As an example, consider the base line, which shows the composition of mixtures of BaO and CuO. As the composition is varied along this base line, two stable compounds, Ba_2CuO_3 and $BaCuO_2$, are found to exist. These compounds are plotted in their proper places on the composition line, illustrating the purpose of the diagram. Similarly a stable oxide Y_2BaCuO_5 (labelled 211), containing all three

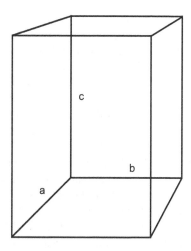

Figure 4.4 The tetragonal structure in this case with the *c* axis longer than the *a* or *b* axes, which are equal in length. The sides are mutually at right angles to each other.

cations, appears on the composition line joining $YO_{1.5}$ and $BaCuO_2$. Following the discovery that a material in this system was superconducting, it was necessary to determine which of several stable phases is the one responsible. Initially, it was believed that a green colored phase, now known to have the composition Y_2BaCuO_5, was the one giving rise to superconductivity. However, it was soon appreciated that this was not in fact the case and that the actual phase responsible for superconductivity is a black one and orthorhombic in structure with a composition $YBa_2Cu_3O_{7-x}$. The position of this now intensively studied compound YBCO is marked on the diagram as 123.

Many of the observed properties of YBCO depend upon its microstructure. Rigorous control of manufacturing and processing is essential to produce either "single crystals", generally necessary for basic investigations, or thin films, tapes, wires and polycrystalline materials, required for commercial applications. Large single crystals of YBCO, and other high temperature superconductors, are still extremely difficult to produce. Most of the single crystals that have been grown are small, thin platelets. Nevertheless, initial difficulties in producing YBCO in ceramic form were soon overcome; well-tried recipes exist (like that described above) and it is now a straightforward process to produce polycrystalline samples. These find many uses.

Structure of YBCO

Most high temperature superconductors have a tetragonal or orthorhombic structure. The tetragonal structure, shown in Figure 4.4, can be thought of in the following way. First consider a cube: each side has the same length ($a = b = c$). Now to make a tetragonal structure, take this cube and stretch it (or squash it) in the direction of one of its sides, keeping a square cross-section perpendicular to the direction of the pull. As a result of this procedure, the direction of pull becomes a unique axis (having a length c) of fourfold rotation, which is different in length from the two untouched sides (a and b) of the square (hence $a = b \neq c$). A number of superconducting cuprates have an orthorhombic structure which has another small distortion so that side a no longer equals side b but a, b and c are still at right angles to each other. That is they have the symmetry shown by a paperback book: three sides of unequal length, each at right angles to the other two ($a \neq b \neq c$).

The structure of the orthorhombic, superconducting phase of YBCO is shown in Figure 4.5. The atomic arrangements and interatomic distances, as determined by x-ray diffraction techniques, show that the structure is related to that of the perovskites ABX_3 (Figure 2.1). It seems natural to place the larger Y^{3+} and Ba^{2+} ions on the A sites, which triples the height of the unit cell, and the smaller Cu ions on the B sites. The tripling of the perovskite unit cell results from arrangement of the Y^{3+} and Ba^{2+} ions such that the top and bottom cells of the stack contain Ba^{2+} ions while the middle cell contains the Y^{3+} ions, as can be seen from Figure 4.5. The arrangement of the oxygen atoms is less easy to determine. There are three anions X per unit cell in the ideal perovskite structure; so in an ideal tripled perovskite unit cell there would be nine oxygen sites available. The fundamental law is that a crystal must have charge neutrality, that is the number of positive and negative charges must balance: seven oxygen ions are needed. $YBa_2Cu_3O_7$ can be thought of as being oxygen deficient relative to the ideal perovskite structure, which would require an oxygen composition of O_9.

How are these oxygen ions arranged? X-ray diffraction experiments show that the yttrium ions are surrounded by only eight oxygen ions rather than the twelve, which would be expected for an ideal perovskite (Figure 3.1). In addition, further oxygen deficiency is observed in the basal planes between the barium ions. However x-ray diffraction studies are not sensitive enough to allow determination of the absolute positions of all the oxygen ions. This is because oxygen is a light atom and scatters x-rays much less strongly than either barium or yttrium. The information required to produce a much clearer picture emerges from neutron scattering studies since neutrons are scattered more effectively by oxygen atoms than are x-rays. Neutron diffraction investigations show that in the basal plane between the barium atoms, one of the oxygen atoms in the cell edge is occupied while the other is not. The cell with the occupied oxygen site (along the b axis of the structure shown in Figure 4.5) is lengthened relative to the cell edge with the vacant site (along the a axis); this produces the orthorhombic structure with b greater than a, although a and b differ by less than 2%.

An important crystallographic feature of YBCO and the other high temperature superconductors is that they are layer-like compounds. As found for $La_{2-x}Ba_xCuO_4$, superconductivity is almost entirely confined within nearly two-dimensional CuO_2 layers. In the case of YBCO there are two adjacent CuO_2 planes only 3.2Å apart separated by yttrium atoms, which lie parallel to each other and are stacked perpendicular to the c-axis (Figure 4.5). We shall see shortly that this means that YBCO is an $n = 2$ compound. Each pair of CuO_2 planes is separated (by planes of other ions)

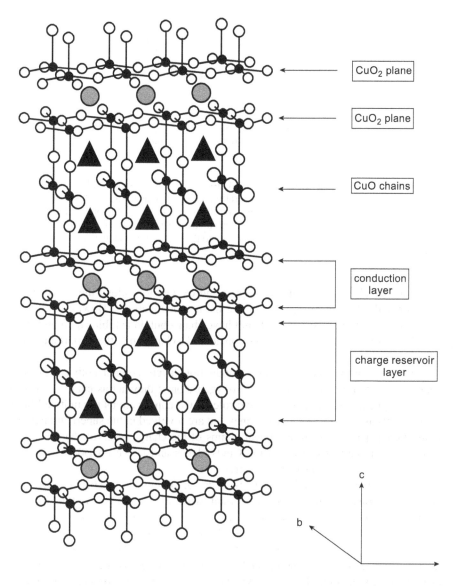

Figure 4.5 Crystal structure of the particularly well known and useful high temperature superconductor $YBa_2Cu_3O_{7-x}$. Small filled circles: Cu; small open circles: O; large shaded circles: Y; shaded triangles: Ba. Superconductivity is almost entirely restricted to the CuO_2 conduction layers, which are puckered.

by the long distance of about 8.2Å from the next pair. The occurrence of superconductivity stems from the fact that each copper ion in the basal CuO_2 layer of the structure is at the center of a square configuration of oxygen ions (Figure 4.1), although the CuO_2 layers are puckered, as can be seen in Figure 4.5. Immediately following the discovery of YBCO, there was a great deal of discussion of the fact

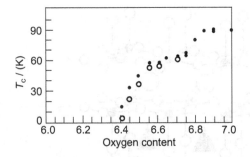

Figure 4.6 An important feature of $YBa_2Cu_3O_{7-x}$ is that its superconducting transition temperature depends on its oxygen concentration shown here as $(7-x)$. In this case x ranges between 1.0 and 0 reading from left to right on the abscissa. The superconducting transition temperature T_c shown by open circles was determined as that at which the electrical resistance reached zero; that denoted by filled circles refers to the onset of diamagnetism. (After Beyers and Shaw (1989).)

that there are also linear chains of copper and oxygen ions along the b axis of the structure and whether they played a role in superconductivity. However the other cuprates with different structures do not have chains but do exhibit superconductivity. Therefore, the chains are not now considered to have quite the importance once ascribed to them.

The regions between the CuO_2 conduction layers contain both barium and copper ions co-ordinated to additional oxygen atoms; they act as charge reservoir layers. It appears that copper plays two roles. Two copper atoms per unit cell lie in the conduction layer i.e. the CuO_2 planes and are involved in the superconductivity. A third copper atom lies in the linear chains of the charge reservoir layers, which control the amount of charge on the CuO_2 planes and hence provide carriers for electrical conduction and superconductivity.

The oxygen content in $YBa_2Cu_3O_{7-x}$ can vary over the range between $O_{6.5}$ and O_7. At the lower limit ($O_{6.5}$), the crystal structure is tetragonal; this form of $YBa_2Cu_3O_{6.5}$ does not go superconducting. To create a superconductor, more oxygen than $O_{6.5}$ is needed. The superconducting form has a composition corresponding to $YBa_2Cu_3O_{7-x}$, where x is less than 0.5 (so that the compound contains more oxygen than designated by $O_{6.5}$). The upper limit is O_7 giving $YBa_2Cu_3O_7$, which has an orthorhombic structure. The exact oxygen content, structure and properties of YBCO depend markedly on the method of preparation. On cooling during its manufacture, it undergoes a second order phase transition at around 700°C to produce a low temperature orthorhombic phase, the superconducting one. The oxygen sites associated with this form are those in the linear chains. There have been extensive investigations into how the critical temperature T_c varies with the oxygen concentration. It has been found that T_c does not vary monotonically with concentration (Figure 4.6). The maximum value is about 93K for a fully oxygenated samples (O_7) and this T_c is retained reasonably well down to around $O_{6.8}$. Between $O_{6.5}$ and $O_{6.7}$ there is a "plateau" with T_c about 60K. Below this concentration, T_c drops rapidly and the material becomes an insulator at $O_{6.4}$. In a series of elegant experiments using neutron powder diffraction techniques, Robert Cava and his group, at AT&T Bell Laboratories, have been able to measure all the Cu–O distances in YBCO as a function of the oxygen loss x from the Cu–O chains. They have correlated changes in the bond length with the valence state of the copper in the CuO_2 planes and observed that it mirrors the changes observed in T_c.

Following the realization of the importance of YBCO, the search was on for closely related compounds. Chu and his colleagues made a further important contribution when they showed that the rare earth element yttrium is not an essential constituent of the new superconductor. When yttrium was substituted by rare earth elements such as lanthanum (La), neodymium (Nd), samarium (Sm), europium (Eu), gadolinium (Gd), holmium (Ho), erbium (Er), and lutecium (Lu), a whole range of superconductors was found which had the same crystal structure as YBCO. The fact that the gadolinium compound is also superconducting was especially interesting because the presence of this strongly magnetic element normally destroys superconductivity.

More Layer-like Cuprate Superconductors

The unique structural feature that the cuprate superconductors contain CuO_2 layer planes was recognized at an early stage and became a guiding principle for the search for similar materials to YBCO. An important step was the discovery by the French group led by Bernard Raveau at Caen of superconductivity at 20K in $Bi_2Sr_2CuO_6$. At first this did not evoke such an immediate response as that generated by YBCO. Yet this modest beginning led to the first 100K superconductors. $Bi_2Sr_2CuO_6$ (often labelled by the acronym Bi2201) can be thought of as the first of a series of $Bi_2Sr_2Ca_{n-1}Cu_nO_{2n+4}$ compounds with n equal to 1. These materials are known familiarly by the acronym BSCCO. The bismuth compounds have layered structures in which CuO_2 layers, typical of all the high temperature superconductors, are spaced by alkaline earth cations, and interleaved with Bi_2O_2 layers (Figure 4.7). The next compound in this series has n equal to 2 so that its formula is $Bi_2Sr_2CaCu_2O_8$. This compound, referred to as Bi2212, has a T_c of 85K. The next one, $Bi_2Sr_2Ca_2Cu_3O_{10}$ (Bi2223) in the series, has n equal to 3 and this layer compound has an even higher T_c of 110K. A trend is clear: the larger n, the higher T_c. But what is the effect on the structure of the key parameter n? It determines the number of immediately adjacent CuO_2 planes in which the superconductivity takes place. A value of unity for n results in the compound with single CuO_2 planes: $Bi_2Sr_2CuO_6$. Pairs of adjacent CuO_2 planes are found in compounds having n equal to two, as in $Bi_2Sr_2CaCu_2O_8$ (and also $YBa_2Cu_3O_{7-x}$). When n is three, the CuO_2 planes occur in sets of three: $Bi_2Sr_2Ca_2Cu_3O_{10}$. The n CuO_2 planes are always perpendicular to the c axis: they are parallel to the $a–b$ plane. Figure 4.7 shows the structure of $Bi_2Sr_2CaCu_2O_8$ (i.e. Bi2212) taken from a review article by the French group at Caen led by Bernard Raveau. It can be seen that Bi2212 has two adjacent parallel planar Bi–O sheets. Between each of the pairs are the perovskite-like multi-layers. These consist of two copper–oxygen sheets in the form of CuO pyramids

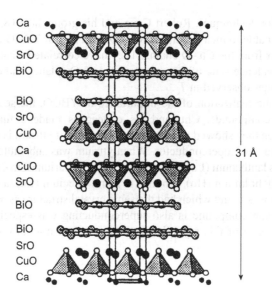

Ca
CuO
SrO
BiO

BiO
SrO
CuO
Ca   31 Å
CuO
SrO

BiO

BiO
SrO
CuO
Ca

Figure 4.7   The crystal structure of $Bi_2CaSr_2Cu_2O_8$ (commonly called Bi2212) is layer-like and includes CuO_2 planes (Figure 4.1) like those of many other high temperature superconductors. (After Bordet *et al.* (1989).)

separated on the base sides by calcium ions. There are no CuO chains; depletion of oxygen leading to insulating behavior is not the problem that it is in YBCO. This can be an important advantage of the bismuth containing compounds over YBCO for technological applications.

Another reason why BSCCO superconductors can be more attractive than YBCO for large scale technological applications is the presence of the weakly bonded, double BiO layer that separates the CuO_2 planes in which superconductivity takes place. These weakly bonded BiO layers give rise to "micaceous" or clay-like mechanical properties. The differences in interlayer bonding between BSCCO and YBCO can literally be felt when one tries to grind the ceramics using a mortar and pestle. Because of the extra weakly bonded Bi–O layers, BSCCO has a "greasy" graphite-like feel to it. By contrast YBCO feels very granular and is rather like sand to grind. This property of BSCCO is an asset in wire making since it tends to produce self-alignment of the grains, assisting a higher current density. A disadvantage of the BiO layers is that they tend to decouple neighboring CuO_2 planes and hence make the BSCCO superconductors strongly two-dimensional.

BSCCO superconductors hold out a great deal of promise for technological development. Quite often some of the bismuth is replaced by lead. This has little effect on the superconducting behavior while helping to stabilize the material. Bi2223 in which some of the bismuth has been replaced by lead is considerably easier to make than when lead is absent.

In general the critical temperature T_c increases through a cuprate series, if the number of neighboring CuO_2 layers is increased from one to two to three. Once this way to proceed in the race towards higher T_c became apparent, many workers attempted to add to or devise new series – with much success. Most of the high temperature cuprate superconductors are now known to fall into such families (Table 4.1). The spacing metal ions include lanthanum (La), barium (Ba), bismuth (Bi), thallium (Tl) or mercury (Hg). The search for other CuO_2 layer plane cuprates led to the discovery of systems of Tl–Ba–Ca–Cu–O superconducting materials. A severe drawback is that thallium is an extremely toxic element. Compounds in this series

Table 4.1 A list of some well known high temperature superconductors. The ideal chemical formula is given together with the superconducting transition temperature T_c. The parameter n refers to the number of immediately adjacent CuO_2 planes. Some other notations that frequently appear in the literature are also given in columns 4 and 5.

Formula	T_c (K)	n		Notations
$(La_{2-x}Sr_x)CuO_4$	38	1	La ($n=1$)	214
$(La_{2-x}Sr_x)CaCu_2O_6$	60	2	La ($n=2$)	---
$Tl_2Ba_2CuO_6$	0–80	1	2-Tl ($n=1$)	Tl2201
$Tl_2Ba_2CaCu_2O_8$	108	2	2-Tl ($n=2$)	Tl2212
$Tl_2Ba_2Ca_2Cu_3O_{10}$	125	3	2-Tl ($n=3$)	Tl2223
$Bi_2Sr_2CuO_6$	0–20	1	2-Bi ($n=1$)	Bi2201
$Bi_2Sr_2CaCu_2O_8$	85	2	2-Bi ($n=2$)	Bi2212
$Bi_2Sr_2Ca_2Cu_3O_{10}$	110	3	2-Bi ($n=3$)	Bi2223
$YBa_2Cu_3O_7$	92	2	Y123	YBCO
$YBa_2Cu_4O_8$	80	2	Y124	-----
$Y_2Ba_4Cu_7O_{14}$	40	2	Y247	-----
$TlBa_2CuO_5$	0–50	1	1-Tl ($n=1$)	Tl1201
$TlBa_2CaCu_2O_7$	80	2	1-Tl ($n=2$)	Tl1212
$TlBa_2Ca_2Cu_3O_9$	110	3	1-Tl ($n=3$)	Tl1223
$TlBa_2Ca_3Cu_4O_{11}$	122	4	1-Tl ($n=4$)	Tl1234
$HgBa_2CuO_{4+\delta}$	94	1	1-Hg ($n=1$)	Hg1201
$HgBa_2CaCu_2O_{6+\delta}$	127	2	1-Hg ($n=2$)	Hg1212
$HgBa_2Ca_2Cu_3O_{8+\delta}$	133	3	1-Hg ($n=3$)	Hg1223

with n equal to 1, 2 and 3 are now quite well studied. Materials with higher values of n are known but they are quite difficult to synthesize and are poorly characterized. The critical temperature T_c rises as n increases up to 3, but then starts to decrease. In the thallium series, as for the bismuth series, when CuO_2 layers are placed adjacent to each other there is a tendency for the interaction, which produces superconductivity to strengthen. But there seems to be a limit to the increase of T_c induced by increasing n.

The next landmark in the field of high temperature superconductivity occurred with the synthesis of mercury-based high T_c cuprates in 1993. The materials belonging to this family exhibit superconductivity at the highest temperature known to date. The critical temperatures for some members of the mercury family are listed in Table 4.1; again T_c increases with n. Strikingly even for n equal to unity, T_c is as high as 94K; comparative values are much lower for the bismuth and thallium $n=1$ compounds. The record T_c at ambient pressure at the time of writing is 133K found for $HgBa_2Ca_2Cu_3O_{8+\delta}$, denoted by Hg1223; it can be raised even further to above 150K by the application of external pressure. It is therefore not surprising that no sooner than these mercury cuprates were discovered than they became the focus of an intense study of physicists, chemists, and materials engineers.

The effort to find other CuO_2 layer series goes on in many laboratories. As these materials become more complex, it becomes increasingly difficult to make them and to optimize their superconducting properties. More than 50 superconducting cuprates have now been discovered. All are layer-like materials sharing the common structural feature of lightly doped CuO_2 planes.

5 A Diversion: Quantum Mechanics, Atoms and the Free Electron Theory of Metals

Up until now, our discussion has been centered on the discovery and basic properties of conventional and high temperature superconductors. One of our aims is to give a simplified account of current understanding of the mechanisms giving rise to superconductivity; however, by its very nature this has to be rather theoretical in content. To set up an explanation for the origins of superconductivity, it is necessary to digress and provide some essential background material on quantum mechanics and explain how it has completely transformed our understanding of the behavior of electrons in atoms and normal metals. This is the purpose of the present chapter, which is therefore a diversion from the main theme. Many of the ideas described originated during the first thirty years of the twentieth century, one of the most exciting and important eras in the history of physics. The innovations and discoveries made during that time mark the transition from classical to modern physics.

The Quantum Mechanical Behavior of Atoms

The concept of the atom as the building block of matter and speculations as to its nature have fascinated scientists and philosophers since the time of the ancient Greeks. However, only during the last hundred years has real progress been made in determining how atoms are put together. Throughout the history of the modern development of atomic theory, great importance has been attached to experimental spectroscopic data and their representation by accurate numerical formulae. Spectral lines are nature's visual signature of the behavior of electrons in atoms. It was the two eminent nineteenth century German scientists Robert Bunsen and Gustav Kirchhoff, working in Heidelberg, who in 1859–60 established the principles of spectral analysis. In particular they showed that each element has a characteristic set of wavelengths at which it can absorb or emit radiation. Their work led them to discover the elements caesium and rubidium and also gave rise to considerable, although not very fruitful, efforts to explain how the spectral wavelengths could be related to the properties of atoms and molecules.

An important breakthrough was made in 1885 by Johannes Balmer, a geometry teacher at a girl's secondary school at Basle in Switzerland. He showed that the wavelengths of the lines of the emission spectrum of the hydrogen atom, which had previously been measured by the Swedish scientist Anders Ångstrom, are related by a simple formula involving the difference between the reciprocals of the square of two numbers. Nearly thirty years later, the significance of Balmer's formula was recognized by the Danish physicist Niels Bohr.

Bohr constructed an ingenious theory of the atom based on a mixture of classical and quantum physics. Much of the groundwork for his theory was carried out while Bohr was at the Physical Laboratories of the University of Manchester in the period 1912–13 during which time the director of the laboratory was Ernest Rutherford. Bohr and Rutherford were close friends and engaged in lively discussions of physics with each other, although they never published a joint paper. Bohr believed firmly in the nuclear atom proposed by Rutherford and appreciated that this model had profound implications. According to classical physics, Rutherford's atom would be mechanically unstable because the electrons orbiting around the nucleus should radiate energy and quickly spiral round and crash into the nucleus. At first, Bohr was interested in understanding the stability of atoms and the arrangement of the electrons within atoms, which gives rise to the Periodic Table of the elements. He believed that understanding spectral lines was too difficult a problem to solve, although in fact that turned out to be the way to advance. By early 1913, Bohr was having considerable problems with his theoretical ideas and was making little progress. But around this time a colleague at Copenhagen, Hans Hansen, who was a skilled spectroscopist, drew Bohr's attention to the simple relationship for spectral lines in hydrogen that had been discovered years earlier by Balmer. For Bohr, this was a revelation and he spent the next few months in frantic activity, which culminated in his paper "On the Constitution of Atoms and Molecules". That is the first of his trilogy of seminal papers, all published in 1913, containing his model of the atom and explanation of the spectral lines of hydrogen, and is best remembered today.

In Bohr's theory of the atom, the electrons follow orbits around the nucleus in accord with Newtonian mechanics in a similar way to the motion of planets around the sun. However, of all the possible orbits, only a restricted few are allowed (Figure 5.1); these are called stationary states. Furthermore, the electron does not radiate energy while orbiting in one of these stationary states. Although the motion in the orbit is described by classical theory, the transition from one orbit to another, resulting in the emission or absorption of radiation, occurs in "quantum jumps", which cannot be understood in classical terms. The frequency ν of the spectral line is given by the famous quantum formula of Max Planck:

$$E_2 - E_1 = h\nu.$$

Here h is Planck's constant ($h = 6.63 \times 10^{-34}$ Joule seconds). E_2 and E_1 are two energy levels with E_2 being greater than E_1 and the equation is concerned with their difference, which is the energy involved when an electron jumps from one level to

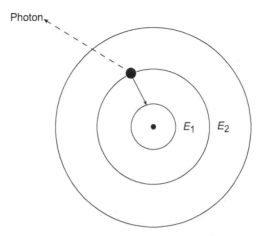

Figure 5.1 The model used by Bohr to interpret the spectrum of the hydrogen atom. Each circle represents an electron state; the larger the radius, the higher the energy E of an electron in the state. When an electron falls down into a lower state, a packet of light (a photon) of specific energy is emitted. The three lowest energy states are shown, each orbit being designated by a principal quantum number n equal to 1, 2, 3 etc.

the other. An electron in the upper, excited state of energy E_2 falls back into a lower energy level E_1 and the energy it loses is given out as a packet of light, as illustrated in Figure 5.1. This energy released gives a spectral line with a definite frequency. Although Bohr had been unaware of Balmer's equation for the spectral lines, as soon as Hansen pointed it out to him, he appreciated that it followed from his concept of an electron changing its state between two quantum orbits and was in excellent agreement with his theory. Bohr fully realized the radical departure, which his introduction of quantization of energy into the theory of the atom was making from accepted ideas in classical physics, something that Planck, who had first introduced the idea of quantized energy in 1900, had been unable to countenance.

More Sophisticated Ideas Are Needed

By the 1920s it had become clear that Bohr's theory was running into serious difficulties: it could only account for a very simple spectrum like that of hydrogen. However, during this decade, several revolutionary ideas in quantum theory were put forward which transformed completely the understanding of the nature of atoms. These ideas centered round Niels Bohr and a stream of exceptionally gifted physicists who visited his Institute for Theoretical Physics in Copenhagen. One of the first of these visitors was Werner Heisenberg who for a while held a lectureship in theoretical physics at Bohr's Institute. Heisenberg argued that the drawback in the Bohr theory was that it was constructed in terms of the position, orbit and velocity of an electron moving around the nucleus and that none of these parameters was observable. He

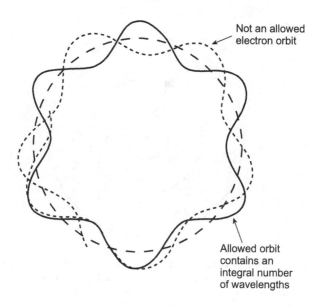

Not an allowed
electron orbit

Allowed orbit
contains an
integral number
of wavelengths

Figure 5.2 In an extension of the Bohr model of the atom the allowed electron orbits are those which fit exactly an integral number of de Broglie wavelengths around the circular orbit. Here the solid closed orbit shows such a standing wave for the *n* equal to 6 level so that 6 complete wavelengths fit exactly round the orbit. The dotted open orbit does not satisfy the Bohr conditions and so is not an allowed state for an electron.

suggested that it would be better to construct a theory in terms of observable quantities such as the frequencies, intensities and polarizations of the radiation emitted by the atoms. Heisenberg developed a theory based on this approach, the details of which he worked out in collaboration with Max Born and Pascual Jordan from Göttingen. It quickly became apparent that the rules, which Heisenberg had introduced for combining the different transition amplitudes in the atom, were the same as the rules of matrix multiplication, well known to mathematicians. Matrix mechanics, as Heisenberg's approach became known, proved to be highly successful and provided valuable insight into understanding light spectra produced by excited atoms.

Around this time, a young Frenchman Louis de Broglie put forward a highly novel unification of matter and waves. In his doctoral thesis, submitted in 1924, he advanced the original idea that the same dualism of wave and particle shown by light might also be true for matter. Just as the light quantum or photon has a light wave associated with it, giving rise to interference and diffraction effects, so de Broglie suggested that a material particle might also have a corresponding matter wave. In addition, the matter wave could be expected to obey similar mathematical equations; hence he postulated that the wavelength λ of the matter wave could be related to the particle momentum p by putting wavelength λ equal to Planck's constant h divided by momentum: λ is equal to h/p. Although at the time many physicists were sceptical about de Broglie's ideas, no other doctoral thesis has ever had such a dramatic impact on our understanding of the physical behavior of the world. It soon became clear that matter waves were of great importance in understanding the nature of the atom. De

Broglie was a connoisseur of chamber music and this led him to visualize the atom as a musical instrument that could emit a basic tone and a sequence of overtones. An important result stemming from his ideas was that he could readily obtain the quantization rules of Bohr, which until then had seemed completely arbitrary, by requiring that an integer number of waves should be fitted around a stationary orbit as shown in Figure 5.2. Within a short while of de Broglie's suggestion of the existence of matter waves, Clinton Davisson and Lester Germer in the United States and George Paget Thomson in Britain independently provided an experimental verification that electrons do indeed have a wave nature. They observed diffraction effects caused by the passage of electrons through metals and were able to establish that moving electrons have associated waves with a wavelength λ given by h/p, that is Planck's constant divided by their momentum, as de Broglie had predicted. Nowadays, the technique of electron diffraction, which utilizes the wave nature of matter, has developed into a powerful research tool in the form of the electron microscope. George Thomson was the son of J.J. Thomson and so "J.J." discovered the electron while "G.P." demonstrated its wave properties, both obtaining Nobel Prizes in physics as a result in 1906 and 1937 respectively.

Although de Broglie had put forward the idea of matter waves, he did not develop his concept into a strict mathematical theory. This groundbreaking step was made in 1926 by the Austrian-born physicist, Erwin Schrödinger, who wrote a general equation for de Broglie waves and then proved its validity for various kinds of electron motion. At the time Schrödinger was in charge of theoretical physics at the University of Zurich. Close by was the prestigious Swiss Federal Institute of Technology having among its staff at the time such luminaries as Peter Debye, a highly versatile scientist who obtained the 1936 Nobel Prize for chemistry for his work on molecules, and Hermann Weyl, an eminent mathematician and philosopher. The two institutes held joint colloquia and on one occasion Debye asked Schrödinger to give a seminar on the matter wave hypothesis of de Broglie, which was beginning to attract widespread attention. Schrödinger did so, but Debye was not impressed with his elegant account of de Broglie's work and remarked that, as a student of the great German theoretician Arnold Sommerfeld, he had been taught that to deal properly with waves one needs a wave equation. Apparently, this casually made remark must have registered with Schrödinger, who shortly afterwards produced his famous wave equation. This had an immediate impact. Schrödinger was quickly able to apply his equation to the hydrogen atom and obtain the same energy levels as those found in Bohr's model of the atom. More importantly, the wave equation can also be applied successfully in much more complex situations. Indeed, the Schrödinger equation is arguably the most important relation in twentieth century physics since, apart from relativistic corrections, it is considered to determine the behavior of all material systems, including superconductivity. Paul Dirac is famously supposed to have said that the Schrödinger equation can explain "all of chemistry and most of physics".

In later work Schrödinger was able to show the equivalence between his wave mechanics and the matrix mechanics of Heisenberg. It was not long before Schrödinger's work was heard of in Copenhagen. Bohr was very impressed with it and quick to recognize its significance. Ever since his seminal paper of 1913, Bohr had been struggling with how to understand quantum mechanics and this had become his chief preoccupation. The interpretation of quantum mechanics was a subject that Bohr discussed frequently with Heisenberg. Bohr felt that his own concepts had been strengthened by the development of Schrödinger's mathematical formalism and that from now on the quantum behavior of matter could be treated with equal validity by either Schrödinger's or Heisenberg's approaches. Heisenberg did not share Bohr's enthusiasm for the rival approach used in Schrödinger's work, regarding it as little more than a useful tool for solving mathematical problems in quantum mechanics, while Bohr believed that it provided new insight into the wave particle duality of matter, which he considered to be central to interpreting quantum theory.

Bohr invited Schrödinger to visit Copenhagen during September 1926 to lecture on his work in wave mechanics but also to stay on to discuss the physical interpretation of quantum theory. Bohr was a friendly and humane man but tenacious and persistent in his aims. He met Schrödinger at the train station and immediately began his discussions. Since Schrödinger was staying at Bohr's home there was no respite and the intense talks began from early morning to late at night with Heisenberg also contributing. Eventually, in exasperation, Schrödinger said: "If all this damned quantum jumping were really here to stay, I should be sorry that I ever got involved with quantum theory". To this, Bohr replied: "But the rest of us are extremely grateful that you did; your wave mechanics has contributed so much to mathematical clarity and simplicity that it represents a gigantic advance over all previous forms of quantum mechanics". Shortly afterwards Schrödinger fell ill, exhausted by his efforts over the last few months. He had to retire to bed and was nursed by the kindly Mrs Bohr. Even so, Bohr spent a lot of time sitting at the end of his bed arguing points.

A major problem associated with the Schrödinger wave theory is that it does not address what it is that is vibrating. It is definitely not the electron, which could be thought of as a point particle analogous to a billiard ball, as this implies the classical idea of a particle with spatial position and a particular velocity, both of which change in a deterministic way from one instant to the next. Instead, the electron itself is in some obscure sense the wave. If one considers electromagnetic waves, it is clear that the vibrating quantities are the electric and magnetic field intensities. By contrast, for Schrödinger's electron waves, one can only say that the vibrational nature of matter is taken into account by the wave function that can be determined from his equation. At first, physicists tried to interpret the wave function in classical terms. A fundamental advance in understanding was made by Max Born, in 1926, when he proposed that the absolute square of the magnitude of the wave function represents the probability that an electron follows a particular path through space. With Born's interpretation, the positions and velocities of each electron are basically random, and they have only certain probabilities of exhibiting particular values. For example, if the amplitude of the wave is zero at some point this means that the probability of finding the electron at that position is also zero.

It was while he was at Copenhagen that Heisenberg put forward his famous uncertainty principle, which also has a fundamental role in the foundations of quantum mechanics. The uncertainty principle states that there are certain pairs of physical parameters, which cannot be measured simultaneously to an arbitrarily high degree of accuracy. If, for instance, one tries to determine both the position x and the momentum p of an electron at the same time, then the product $(\Delta x \cdot \Delta p)$ of the average uncertainties (or errors of measurement) of position Δx and momentum Δp must be greater than $h/2\pi$. The closer the position of an electron is defined, the less well known its momentum. Again Planck's constant h shows its deep fundamental significance. Heisenberg regarded the uncertainty principle as central to his own interpretation of quantum theory. He had many long and profound discussions with Bohr over this issue. Bohr realized that both wave and particle ideas are required in order to understand the behavior of any system on the atomic scale. Although both are necessary, they are mutually exclusive. To account for this, Bohr introduced the concept of complementarity: when an atomic system is observed under different experimental conditions, it cannot be understood within the framework of a single model. The wave model of an electron is complementary to the particle model. Bohr realized that whenever one spoke about atomic systems, one always had to resort to using the ideas of classical physics like wave and particle. Such terms are inadequate when describing a system that is so far removed from the classical situation. The principle of complementarity is central to what became known as the Copenhagen interpretation of quantum mechanics. Although at first Bohr was sceptical about Heisenberg's ideas, he later regarded the uncertainty principle as the price one had to pay when applying two concepts that are mutually exclusive to describe atomic systems.

Another frequent visitor to Copenhagen was Wolfgang Pauli who was born in Vienna in 1900, the son of the professor of colloidal chemistry at the University. From an early age Pauli showed a prodigious academic talent for science and mathematics, managing surreptitiously to read and understand Einstein's theory of general relativity during rather boring school lessons. Pauli studied theoretical physics at Munich under Sommerfeld at whose request he wrote a substantial monograph on relativity theory for an encyclopedia on mathematical physics. This monograph is still regarded as a masterpiece to this day. Later, he spent a year in Copenhagen with Bohr. In 1925, remarkable insights by Pauli led to a much better understanding of the behavior of electrons in atoms. He was interested in the alkaline earth elements, such as calcium, and trying to understand and formulate the rules governing the complex series of sharp spectral lines emitted as electrons dropped down spontaneously from a higher energy level to a lower one in atoms, as in the Bohr theory. He was able to formulate a simple underlying concept, now recognized as one of the fundamental ideas of physics, known as the Pauli exclusion principle. His idea was that in any atom no two electrons can have the same set of quantum numbers; each electron state in an atom cannot be occupied by more than one electron. This is because two electrons cannot be in the same place at the same time. In his work on spectra, Pauli (and others) found that it seemed necessary to introduce an additional quantum number, which could only take two values. The interpretation of this additional quantum number also came in 1925, when George Uhlenbeck and Sam Goudsmit suggested that an electron can spin in one of two possible directions (the electron is referred to colloquially as being either spin up or spin down).

The introduction of the two fundamental concepts of the exclusion principle and electron spin led to an understanding of how electrons in atoms are arranged in a series of shells of increasing energy because each orbit is restricted to a maximum of two electrons (one spin up and the other spin down). This ladder of shells neatly accounts for the periodic table of elements, and in addition, the valence states of the elements and hence their chemistry. For several years prior to the work of Pauli, Bohr and his colleagues at Copenhagen had struggled to understand the periodic table. They had achieved quite a measure of success as can be seen by the discovery of element number 72, which Bohr predicted would have properties similar to the element zirconium (Zr) rather than being a rare earth element as was widely thought at that time. Analysis of zirconium rich minerals by Dirk Coster and George de Hevesy did indeed lead to the discovery of element number 72 in Bohr's own laboratory in Copenhagen. The new element was called hafnium (Hf), after the Latin name for Copenhagen, and Bohr announced its discovery at the end of his Nobel prize lecture in 1921. Nevertheless, the work of Pauli placed Bohr's speculative ideas on a much firmer theoretical footing. Elements fall into their positions in the periodic table because of the arrangement of electrons in their atomic shells as determined by the Bohr theory.

The wave nature of electrons, the Pauli exclusion principle and the operation of the uncertainty principle are all essential concepts for an understanding of super-conductivity, which is a purely quantum cooperative effect of electrons acting in concert.

Cooperative Effects: Magnetism, Spin and Superconductivity

In addition to his work on the foundations and philosophy of quantum mechanics, Heisenberg, one of the most versatile of physicists, extended his ideas to evolve a basic understanding of ferromagnetism. Magnetism, like superconductivity, arises from cooperative effects of electrons. The phenomenon has been recognized as far back as the ancient Greeks, who used the lodestone as a primitive compass. Every-body is aware of the powerful magnetic effect of iron in attracting objects like nails. Iron is a common example of a ferromagnetic material along with the other elements cobalt and nickel.

Magnetism exists in a variety of different forms. In Chapter 2, we discussed the work of Faraday, who was able to distinguish between two forms of weak magnetism, namely paramagnetism and diamagnetism depending on whether the material was attracted or repelled by a strong magnetic field. A type I superconductor is an example of a perfect diamagnet, since all the magnetic flux is suddenly expelled at the super-conducting transition. Ferromagnetism and superconductivity are sworn enemies in that the former will always try to destroy the latter. However, recently, some ingenious work has shown that it is possible for the two cooperative properties to coexist in certain exotic materials.

Ferromagnets possess a spontaneous magnetization; that is a magnetization even in the absence of an applied magnetic field. It is a cooperative effect in which all the spins on adjacent atoms line up parallel to each other as illustrated in Figure 5.3(a). By contrast, in antiferromagnets, which show another type of magnetic state, the spins line up cooperatively antiparallel to one another as illustrated in Figure 5.3(b). This antiparallel alignment takes place below a characteristic temperature T_N

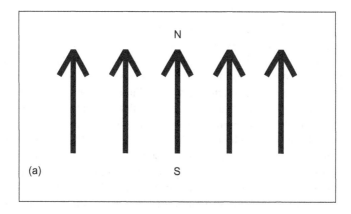

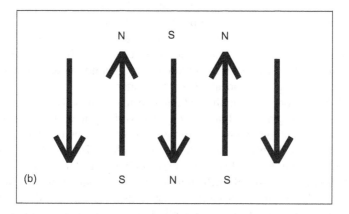

Figure 5.3   Ordered arrangement of spins on the ions in (a) a ferromagnetic material – all the spins are parallel and (b) an antiferromagnetic material – adjacent spins are antiparallel.

generally referred to as the Néel temperature after the distinguished French physicist Louis Néel who pioneered studies of antiferromagnetism and other forms of magnetism for which he shared the Nobel prize for physics in 1970. The parent materials of most of the high temperature superconducting cuprates are antiferromagnets, a feature that has important consequences (see Chapter 11).

Bosons and Fermions

Superconductivity is a property of metallic materials; to understand it further, we need to know how electrons behave in metals. It was in an effort to understand how electrons carry a current in metals that Kamerlingh Onnes himself had made his experiments on the electrical resistance of mercury at low temperature (after previously investigating gold and platinum) and discovered superconductivity (Chapter 1). His work was prompted by conflicting theories of electrical resistance, which he believed careful measurements at low temperatures would resolve.

In 1897, J.J. Thomson had discovered the electron. Remarkably enough by 1900 – only three years later – understanding of the behavior of electrons in metals began with an idea proposed by the German physicist Paul Karl Drude that a metal can be pictured as a box containing a "gas" of freely circulating electrons. The neutral atoms donate electrons to the conducting electron gas and become positive ions. This "free electron" model, further developed by the Dutch physicist Hendrik Lorentz, proved to be very successful in explaining many properties of metals. Most importantly, it accounted for the conduction of electricity by means of the movement of the valence electrons (as distinct from the electrons of the filled shells of the ion cores of the atoms in the metal). It also showed that these conduction electrons are involved in the conduction of heat. However, some of the predictions of the model were not at all accurate. In particular, the specific heat of the electron gas (relating to the amount of energy required to raise the temperature of the gas by a given amount) was predicted to be many times larger than that observed experimentally. Where the theory went wrong is that it assumed that it was necessary to heat all the electrons in the electron gas. The reason for the failure of this assumption did not emerge until the development of quantum ideas a generation later.

In 1926, the modest and apolitical Italian Enrico Fermi and the Englishman Paul Dirac, two of the originators of nuclear physics, working independently, applied the Pauli exclusion principle to an "electron gas", which can be thought of as a large number of isolated electrons each with a spin of either $+1/2$ or $-1/2$. They derived what is known as the Fermi–Dirac statistics and introduced the concept of a Fermi temperature T_F below which the behavior of the "electron gas" is quite different from that which would be expected from classical particles. The statistics derived by Fermi and Dirac are quantum statistics and applicable to particles with spins of 1/2, 3/3, 5/2 ... etc. Particles, such as electrons, which obey Fermi–Dirac statistics are known as fermions. A very different quantum statistics is applicable to particles with a spin of 0, 1, 2, 3... and so on. This second class of particle behavior, pertaining to light and heat and later appreciated to have an important bearing on superconductivity, was first investigated by the Indian physicist Satyendra Bose and extended by Albert Einstein; hence the statistics that such particles obey is known as Bose–Einstein. Particles obeying these Bose–Einstein statistics are called bosons and obey very different laws from fermions. With bosons, the Pauli exclusion principle does not apply: any number of bosons can occupy a particular energy state (that is, they can have precisely the same set of quantum numbers that denote that state). The rules about the different ways in which fermions and bosons occupy energy states and the arrangement of the two kinds of particles induced by them are illustrated in Figure 5.4. They cause the behavior of the two types of quantum entities fermions and bosons to be very different: differences which lie at the heart of the nature of the universe. It is possible at low temperatures for an appreciable fraction of a collection of bosons to be in exactly the same energy state; such an ensemble is known as the "Bose–Einstein condensate". The best-known examples of bosons include the quantized particles of light called photons, superconducting electrons and the

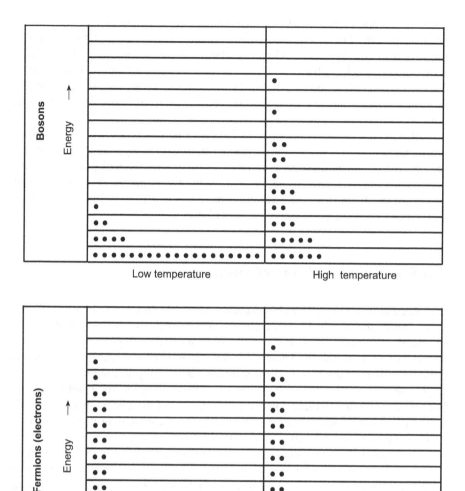

Figure 5.4 Bosons and fermions obey different laws which determine how they arrange themselves in quantum states at any temperature. Any number of bosons (such as photons or phonons) can occupy the same quantum state, so that at low temperature most tend to go into the ground state (as seen in the top left column). By contrast, since fermions (such as electrons) obey the Pauli exclusion principle, only one of them can occupy a given state (bottom two columns). A spin-up electron is in a different quantum state from one with spin-down; that is why two electrons are shown in states here. At low temperatures there is a sharp cut-off (bottom left column), corresponding to the Fermi level in a metal, between electrons in filled and empty states; even at high temperature (bottom right column) most conduction electrons remain in states below this level.

common isotope of helium He^4, which has two protons and two neutrons in its nucleus. At very low temperatures, below 2.18K, atoms in liquid helium collect together in a Bose–Einstein condensate and show the remarkable property known as superfluidity, which is a frictionless motion of atoms with striking parallels to super-conductivity. It will be shown in the next chapter that this condensation can be useful in hand-waving arguments to explain superconductivity (Cooper pairs behave as bosons).

There is a colorful analogy for the difference between fermions and bosons at low temperatures. If one considers a railway train with many individual carriages, then people from the more straight-laced Nordic countries will probably sit in-dependently one to a carriage and can be thought of as fermions. By contrast, jovial and gregarious Latins from the Mediterranean region will all crowd into one carriage and can be thought of as bosons – particles of heat and light!

The Free Electron Theory of Metals

Once quantum mechanics, Pauli's exclusion principle and, most significantly, the Fermi–Dirac statistics had been developed, the way was opened to a realistic theory of metals. However, that entry was not immediately taken. Ralph Fowler, Rutherford's son-in-law and Dirac's doctoral supervisor, who had first appreciated the implications of the Fermi–Dirac statistics for astrophysics, but not for metals, wryly commented later, "I had the theory of electrons in metals right under my nose but I couldn't see it was there. I kick myself whenever I think of it". It was not until 1928 that Sommerfeld realized that the behavior of electrons in metals could be understood by using the Fermi–Dirac statistics to take account of the sequential manner in which electrons occupied the available energy levels. In the free electron gas model of a normal metal, the valence electrons are assumed to be free to move throughout the whole crystal, while the inner shell electrons remain localized on the ion cores. The potential energy of the valence electrons is assumed to be constant within the bulk of the metal, while there is a high potential barrier at the surface, and a large force is required to remove electrons completely from the solid: the electrons stay in the metal box! In these circumstances, the conduction electrons act as travelling waves. The mathematical basis of how electrons behave in a metal is given in Box 1. Note that the material set in each of the boxes provides a more detailed mathematical and physical background but does not alter the line of the argument in the main text. Take it on trust, if you wish.

The energy levels in the electron gas form a closely spaced ladder of increasing energy, as shown diagrammatically in Figure 5.5. The way in which the valence electrons occupy these states is determined by the Pauli exclusion principle: each level can hold only two electrons, one spin up and the other spin down, just as found for atoms. The electrons obey the rules illustrated in Figure 5.4. The effect of the Pauli exclusion principle is to compel electrons to go into the higher energy levels when they find lower levels full. As a result, the electrons in the gas occupy the states from the lowest kinetic energies up through the higher ones until a maximum

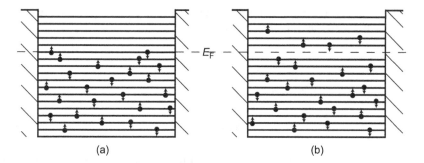

(a)   (b)

Figure 5.5   The Sommerfeld model for a metal as a box containing a gas of valence electrons. Energy increases upwards. (a) At the absolute zero of temperature (0K) all the electrons in the electron gas occupy each energy level in pairs. This filling of the available states continues up to the Fermi level E_F, which separates the filled from the empty states. (b) At temperatures above 0K electrons in levels just below the Fermi level acquire thermal energy and are excited into states just above the Fermi level. (After J.M. Ziman *Electrons in Metals* 1963.)

occupied energy state, known as the Fermi energy E_F, is reached above which the states are empty. The kinetic energy of the electrons, which are at the Fermi level and are thus the most energetic ones in the electron gas, is large: of the order of 5eV in a metal. Indeed this energy is enormous compared with that for thermal excitations at room temperature of 0.025eV (equal to $k_B T$ where k_B is the Boltzmann constant). This large kinetic energy means that electrons at the Fermi level are travelling at about 10^6 meters per second, only a few hundred times slower than the velocity of light. In this sense the electron gas would seem to be extremely hot, with a temperature T_F (equal to E_F/k_B, the Fermi energy E_F divided by the Boltzmann constant k_B) of about 50,000K. However the metal is not hot in the usual meaning of that word: interaction of the electrons in the gas with the ions holds the metal together and the electrons cannot give up any of their energy to a colder body (such as an inquisitive finger!). Even when the temperature of a metal is lowered towards the absolute zero, the electrons still retain the high Fermi energy; the electron system stabilizes itself at the lowest possible energy, that is it drops into its ground state energy. This corresponds to the state in which all the levels below the Fermi energy are filled and all the higher levels are empty.

Box 1

The free electron theory of metals

Solution of the Schrödinger equation treats conduction electrons in a metal as travelling waves defined in one dimension by the wave function:

$$\psi = \exp(i\{kx - \omega t\}). \tag{1}$$

Here the wavevector **k**, in terms of the wavelength λ of the associated de Broglie waves, is $2\pi/\lambda$; time is t and the angular frequency is ω ($= 2\pi\nu$), where ν is the

wave frequency. From the de Broglie hypothesis, the wavelength of waves associated with a particle of momentum $\mathbf{p}$ is:

$$\lambda = h/\mathbf{p}. \tag{2}$$

Therefore, writing $\hbar$ for $h/2\pi$, we have:

$$\hbar\mathbf{k} = \mathbf{p}. \tag{3}$$

Motion of the free electrons is determined by reference to their associated de
Broglie waves. This treatment of free electrons in terms of the wave vector $\mathbf{k}$ can
be readily justified from the Heisenberg uncertainty principle:

$$\Delta x \Delta p \approx \hbar. \tag{4}$$

The uncertainty of the position Δx of a free electron in a crystal lattice is large
and, in consequence, the uncertainty of momentum Δp of the electron is small,
and momentum, or its equivalent $\hbar\mathbf{k}$ from equation (4), becomes useful for the
description of electron motion in a lattice. An electron with a wave-vector $\mathbf{k}$ is
said to be in the state $\mathbf{k}$ and the kinetic energy $E(\mathbf{k})$ of the electron is given by:

$$E(\mathbf{k}) = m\mathbf{v}^2/2 = m^2\mathbf{v}^2/2m = \mathbf{p}^2/2m = \hbar^2\mathbf{k}^2/2m. \tag{5}$$

This equation (of a sphere in $\mathbf{k}$-space) determines the energy of the states
available for the conduction electrons, which comprise the electron gas in the
metal box.

The Fermi Surface

From the physical and mathematical points of view, the natural way of classifying
electronic states in a metal is to plot them on a three-dimensional graph whose axial
scales are the electron momentum, which is equal to $\hbar\mathbf{k}$. Usually $\mathbf{k}$ (the wavevector)
is used instead of momentum $\hbar\mathbf{k}$. Surfaces of constant energy constructed in this
"k-space" are found to be very useful. The constant energy surface that separates the
filled from the empty states is called the "Fermi surface". The concept of the "Fermi
surface", introduced in Box 2, gives a fuller picture of how the electrons behave in a
metal – central to a formal description of the behavior of electrons in metals. It is not
a surface in the real sense but a mathematical concept. Metal physicists speak of
"electrons on the Fermi surface". The name has been traced back to a paper by
Bardeen, in 1940, although the emerging concept can be found in much earlier
papers, as long ago as 1930. The first Fermi surface to be determined experimentally
was that of copper by Brian Pippard in 1957. Electrons near the Fermi surface are the

Box 2

Momentum space or k-space: the standard way of visualizing electrons in metals

A useful way of picturing the behavior of the electrons is to show their momentum position in a plot inside a three-dimensional momentum space. Since the electron momentum $\mathbf{p}$ is given by $\hbar\mathbf{k}$, solid-state physicists usually choose to plot the electron position in $\mathbf{k}$-space, which scales a momentum space down by $\hbar$. An electron contained in a metal box of volume V can only take certain values of $\mathbf{k}$ which are evenly and very closely spaced, when V is large enough to contain many atoms. A simplified two-dimensional diagram of $\mathbf{k}$-space is shown in Figure 5.6.

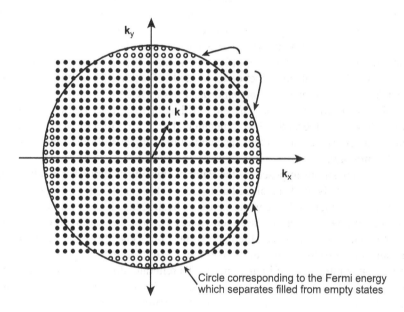

Circle corresponding to the Fermi energy
which separates filled from empty states

Figure 5.6   The way in which the available electron states are arranged in "$\mathbf{k}$-space" in a two dimensional model. The states form an array of equally spaced points each of which can contain two electrons, one spin up, the other spin down. For free electrons the surfaces of constant energy are circles; therefore the states at the corners of the square are higher in energy than those within the circle which corresponds to the Fermi energy – electrons go into the states below that circle rather than into those near the corners of the square. (After J.M. Ziman *Electrons in Metals* 1963.)

Each point represents a state. Only two electrons can occupy each state, i.e. exist at a point. Each $\mathbf{k}$ value represents a state of discrete momentum and energy. This figure illustrates, for two dimensions, how electrons fill up the available electron $\mathbf{k}$-states. Here radii of successive circles centered on the origin represent increasing momentum $\mathbf{p}$, which is equal to $\hbar\mathbf{k}$. The electron energy

increases with increasing **k** and for free electrons, surfaces of constant energy
are circles. For a three dimensional solid they are spheres in three-dimensional
k-space (following equation (5) in Box 1). At the absolute zero of temperature
the gas of free electrons takes the lowest possible energy, by packing the
occupied states in a sphere of radius k_F, centred on the origin. The sphere that
represents the boundary between the filled and empty states is called the "Fermi
surface".

ones which carry electric current in a metal and are those which enable superconductivity to occur.

Since the early days when quantum mechanics was first applied to crystalline
solids, an understanding of the conventional metallic state has been developed in
great detail. This achievement is remarkable, considering that the problem involves
an enormous number of electrons interacting strongly with each other and with
atomic nuclei. A metal can be considered to be a state of matter that has a Fermi
surface and, using this model, its low energy behavior can be understood without
detailed knowledge of the complex interactions that actually occur between the
electrons. A readable introduction to the study of electrons in metals can be found in
articles by J.M. Ziman in *Contemporary Physics* for 1962 which were collected in
book-form as *Electrons in Metals* 1963. Although this was written some 40 years ago,
in the heyday of the study of the fundamental properties of normal metals, it remains
unsurpassed as a clear and lucid account of the subject still referred to nostalgically
as Fermiology.

The constraint imposed by the Pauli exclusion principle that electrons occupy
the energy levels singly resolved the mystery of the comparatively small specific heat
of the electron gas. When a metal sample is heated, the electrons in states well below
the Fermi level cannot increase their energy because the states above them already
contain electrons. However, there are empty states just above the Fermi level. This
means that electrons in states near the Fermi energy, but only these, can be
accelerated into states of higher energy. To reiterate, this results in the arrangement
of electrons in energy states shown in Figure 5.4. At any finite temperature, thermal
energy (which is about 0.025eV at room temperature) is absorbed and some electrons
are excited from states, such as that labelled k_1 in Figure 5.7, just below the Fermi
level into states, such as k_2 just above the Fermi level. The number of occupied states
above the Fermi level is equal to the number of empty states left below it. As a result
of the constraint that only one electron can occupy a given state, when the temperature is increased only a relatively few electrons, those near the Fermi energy, are
able to increase their velocity and momentum by being excited into higher energy
states. This is why the specific heat of the electrons is so much smaller than classical
physics had predicted.

In the case of sodium metal the picture of a spherical Fermi surface works well,
giving a good description of the behavior of the electrons. A two-dimensional cross-
section is shown in Figure 5.6. However the Fermi surfaces of most other metals are
rather exotic objects, leading John Ziman to point out that they might be mistaken
for pieces of modern sculpture. Alan Mackintosh has suggested that although most

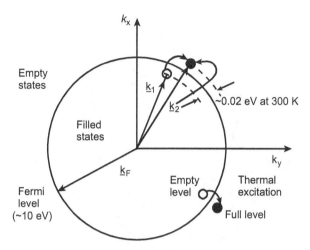

Figure 5.7 In the free electron model of a metal the Fermi surface is a sphere. At 0K all the states inside the sphere contain electrons have energies less than the Fermi energy which is that on the Fermi surface sphere. All states outside the sphere are empty. However at any higher temperature some electrons from states just below the Fermi level are thermally excited into states with energies greater than that at the Fermi surface, leaving some empty states below. The scale is shown by the fact that the Fermi energy in a metal is of the order of 10eV while the thermal energy at room temperature (300K) is only about 0.02eV; so only electrons in a very thin skin at the Fermi surface can be thermally excited.

people would not define a metal as a "solid with a Fermi surface" that might be the most meaningful definition. A detailed understanding of why a metal behaves as it does requires knowledge of its Fermi surface, and in consequence, great efforts have been extended to construct them for most metals. As might be expected, the Fermi surface can help to understand the behavior of a superconductor. That for niobium is shown in Figure 5.8. The electrons, which are the source of superconductivity in a conventional superconductor such as niobium, are those in states close to the Fermi surface.

A central question for understanding the nature of superconductivity in the high T_c cuprates is whether they can also be considered to have a Fermi surface. If so, then a theory of superconductivity in these materials can assume the existence and meaningfulness of a Fermi surface in the normal state. Whether or not this is so, calculations of the Fermi surface can be, and have been, made on the basis of "band theory". A picture of such a calculated Fermi surface is given for YBCO in Figure 5.9. To test the theory, many experiments have been carried out. In general, the results imply that the cuprates can be described as having a Fermi surface and that its form is predicted reasonably well by the band theory. This observation limits the types of theories that need to be considered to explain high temperature super-conductivity. In the simple free electron model of metals electrons act as the carriers of electricity. However as we have seen, the majority carriers of electric current in most cuprates are the "holes", or simplistically the positively charged spaces left by electrons at the top of a nearly filled band.

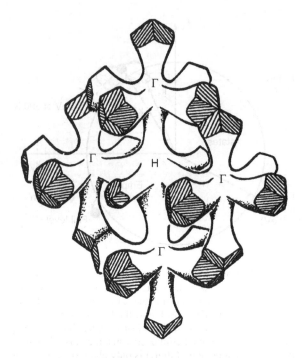

Figure 5.8 One part of the Fermi surface of niobium; there are several other pieces. The letters label points of high symmetry. The Fermi surface is plotted in **k**-space. Niobium has the highest critical superconducting temperature of any element. (After Scott and Springford (1970).)

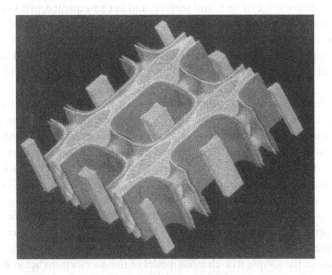

Figure 5.9 A theoretically calculated Fermi surface of YBCO. (After Pickett *et al.* (1992).) Experimental measurements have been made of some pieces and are in reasonable agreement with the predicted surface.

Metals are good conductors of electricity because they contain electrons in states (near the Fermi surface) adjacent in energy to empty states: application of an electric field causes these electrons to be accelerated into the unoccupied states and conduct the current. Details of the conduction mechanism in the free electron model are outlined in Box 3. In normal metals, the source of the resistance is scattering of electrons from one state to another.

Box 3

How metals conduct electricity

For the free electron model, the electron energy depends upon the square of the wave vector **k** (equation (5) in Box 1); this parabolic dependence of the energy on the wave vector **k** is shown in Figure 5.10. Although this simple model does

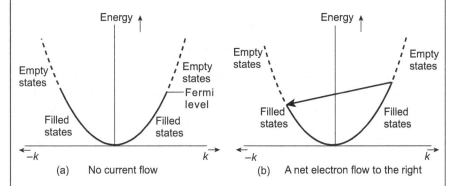

(a)   No current flow   (b)   A net electron flow to the right

Figure 5.10   The mechanism of the conduction of electricity and the electrical resistivity in a one-dimensional normal metal. The full lines represent electrons occupying the available states and the dotted lines represent empty states. (a) When there is no applied electric field, the occupation of available states by electrons is the same for both negative and positive values of **k**; hence the momentum in the negative direction is the same as that in the positive direction and there is no net current flow. (b) When an electric field is applied, electrons move towards the positive electrode and there is an increase in the number of electrons in that direction (in this case corresponding to a positive value of **k** where the level of filled states rises and momentum is increased in that direction). A balance is achieved by scattering of electrons (shown by the long arrow) back from the higher energy levels on the right into lower energy states on the left – this scattering constitutes the electrical resistance.

---

not hold exactly for any real metal, it is a close enough approximation to the true state of affairs to be useful for our later goal of accounting for the absence of resistance in a superconductor. For a normal metal, flow of electric current and electrical resistance can be readily understood. Figure 5.10, drawn using equation (5), is a plot of the electron energy against **k**, which is equal to momentum divided by a constant. If no electric field is applied, the electrons fill the lowest energy states, which are those at the bottom of the parabola. This is shown in Figure 5.10(a). There are as many electrons with negative values of **k** as with positive values, and the total momentum of all the electrons, given by the summation of $\hbar\mathbf{k}$ for each electron, is zero, as it must be if there is to be no net flow of current. However, when an electric field is applied across the metal, a current is induced and the total kinetic energy of the system increases. The electrons move into $\mathbf{k}$-states in such a way, illustrated in Figure 5.10(b), that there is a net momentum along the direction of electron current flow

Electrons near the Fermi level are constantly being scattered into different $\mathbf{k}$-states by collisions with thermal vibrations and lattice defects. Scattered electrons tend to go into the empty states with momentum directed in the opposite direction to the electron current flow (such as those on the LHS in Figure 5.10(b)), because these states are lower in energy than any other available states. Therefore, in an electric field a condition of steady current flow is reached which represents a balance between the energy and momentum increases arising from acceleration of the electrons in the field and the decreases due to scattering back into lower energy states. When the electric field is removed, the scattering processes cause the system to revert to the state of lowest energy, which is illustrated in Figure 5.10(a) and electrical conduction ceases.

In superconductors there is no resistance and the scattering processes, which occur in the normal state, must be inoperative. This happens because the energy states are not arranged in the same way as those for a normal metal. In order to understand the origin of superconductivity in conventional metals, we now need to turn our attention to the physical picture developed by Bardeen, Cooper and Schrieffer, the subject of the next chapter.

6 What Causes Superconductivity?

Following Kamerlingh Onnes' discovery of zero resistance, it took a very long time to understand how superconducting electrons can move without hindrance through a metal. Attempts to explain from first principles how superconductivity comes about proved to be one of the most intractable problems of physics. Progress required more than just new data; it needed an innovative theoretical framework built around radical new ideas. There are inherent difficulties in achieving this. Fundamentally new concepts are not discovered by observation alone but require new modes of imaginative thought, which by their very nature are unpredictable and elusive.

One day in 1955, when John Bardeen, Bernd Matthias and Theodore Geballe were driving between Murray Hill and Princeton, the question was raised: "What are the most important unsolved problems in solid state physics?" After a characteristic lengthy pause, Bardeen suggested that superconductivity must be a candidate. Later in 1957, in collaboration with Leon Cooper and Robert Schrieffer, he was to provide a most ingenious, and generally accepted, explanation of superconductivity founded on quantum mechanics. The physical principles underlying this BCS theory are the concern of this chapter. It has become evident that while the BCS theory gives a reasonable description of superconductivity in the "conventional" superconductors, known at that time, that is not the case for more recently discovered "unconventional" materials, such as the high temperature superconducting cuprates (which will be discussed separately in Chapter 11).

BCS theory has established that superconductivity in conventional materials arises from interactions of the conduction electrons with the vibrations of the atoms. This interaction enables a small net attraction between pairs of electrons. Before insight into this electron pairing and a subsequent ordering can be gained, some characteristics of superconductors need to be brought to mind.

Superconductivity is a common phenomenon; at low temperatures many metals, alloys and compounds are found to show no resistance to flow of an electric current and to exclude magnetic flux completely. When a superconductor is cooled below its critical temperature, its electronic properties are altered appreciably, but no change in the crystal structure is revealed by X-ray crystallographic studies. Furthermore, properties that depend on the thermal vibrations of the atoms remain the same in the superconducting phase as they were in the normal state. Superconductivity is not associated with any marked change in the behavior of the atoms on the crystal lattice. However, although superconductivity is not a property of particular atoms, it does depend on their arrangement. For example white, metallic tin is a superconductor but grey, semiconducting tin is not. Another of the many examples illustrating an involvement of the atomic arrangement is that while the usual semimetallic form of bismuth is not superconducting, even at a temperature as low as 10^{-2}K, several of its crystalline forms, which can be obtained under high pressure, are. Although the atomic arrangement is important, it need not be regular in the crystalline sense; even some glassy materials can become superconducting. For example, bismuth when condensed onto a cold finger surface at 4K forms an amorphous film; in this form it is superconducting. Amorphous metals have recently become very important materials. Alloys can be frozen in an amorphous state by cooling them extremely quickly from the melt; some are superconductors.

The conduction electrons themselves must be responsible for the superconducting behavior. A feature which illustrates an important characteristic of these superconducting electrons is that the transition from the normal to the superconducting state is very sharp: in pure, strain-free single crystals it takes place within a temperature range as small as 10^{-3}K. This could only happen, if the electrons in a superconductor become condensed into a coherent, ordered state, which extends over long distances compared with the distances between the atoms. If this were not so, then any local variations from collective action between the electrons would broaden the transition over a much wider temperature range. A superconductor is more ordered than the normal metal; this means that it has a lower entropy, the parameter that measures the amount of disorder in a system. In an analogous way, the entropy of a solid is lower than that of a liquid at the same temperature; solids are more ordered than liquids. A crucial conclusion follows. When a material goes superconducting, the superconducting electrons must be condensed into an ordered state. To understand how this happens, we need to know how the electrons interact with each other to form this ordered state. That mechanism is the essence of the BCS model.

The Isotope Effect

In the search for the nature of the interaction that binds the electrons together, there was a key question to answer. How do the atoms or their arrangement in a solid assist in the development of superconductivity? An important clue to the form of the way in which they interact with the electrons came from experimental observations in 1950 that the critical temperature T_c depends on the isotopic mass M of the atoms comprising a sample. Many elements can have nuclei having different numbers of neutrons and so have different masses. These isotopes of a given element have identical electronic structure and chemical properties. If the atoms are involved, then changing their mass might be expected to have an effect on superconducting properties. Kamerlingh Onnes himself had looked at this possibility as early as 1922. At that time there were available to him two naturally occurring forms of lead (Pb) having different masses; the more abundant, with an atomic mass M of 207.2, comes from non-radioactive ores, the other from uranium (U) derived lead has a mass M of

206. These two forms of lead differ in mass because they are made from different mixtures of isotopes. In those early days the sensitivity of Onnes' measuring equipment was not good enough to enable him to detect any difference between the superconducting transition temperatures T_c of specimens containing different amounts of lead isotopes. Later experiments made by others, again using lead, were also unable to detect an effect of atomic mass on transition temperature T_c. However with the development of nuclear reactors after World War II, it became possible to make artificial isotopes in sizeable quantities and in turn samples having wider mass differences; at last experiments could be carried out which were able to detect the effect of the atomic mass on transition temperature T_c. The required high sensitivity of measuring temperature is illustrated by the fact that for mercury T_c varies only from 4.185K to 4.146K as the atomic mass is changed from 199.5 to 203.4.

The measurements made on mercury samples having different isotopic masses, and later on lead and tin, showed that the superconducting transition temperature T_c is proportional to the inverse square root of the atomic mass M (that is T_c is equal to a constant divided by $\sqrt{M}$). Therefore the critical temperature of a sample composed of lighter isotopes is higher than that for a sample of heavier isotopic mass. Changing the isotopic mass alters neither the number nor the configuration of the orbital electrons. This *isotope effect* shows that the critical temperature T_c does depend upon the mass of the nuclei, and so the vibrating atoms must be involved directly in the mechanism which causes superconductivity. The argument is greatly strengthened by the fact that the frequency of atomic vibrations in a solid is also inversely proportional to the square root ($\sqrt{M}$) of the nuclear mass: this correlation strongly suggests that the lattice vibrations must play an important role in the process leading to the formation of the superconducting state.

It is firmly established that the electron–lattice interaction plays a central role in the mechanism of superconductivity in conventional materials. Now at low temperatures, the lattice vibrations, which carry heat or sound, are quantized into discrete energy packets called phonons (from the Greek *phonos* for sound). It is usual to talk about the electron–phonon interaction. In 1950, Herbert Fröhlich from Liverpool University, yet another émigré from Nazi Germany, first tried to produce a theory of the superconducting state based on electron–phonon interactions, which yielded the isotope effect but failed to predict other superconducting properties. Fröhlich realized that electron–phonon interactions could explain the paradox that those elements that are the best conductors of electricity (copper, silver and gold) do not become superconductors even at temperatures as low as 10^{-3} while poorer conductors like lead (T_c = 7.2K) and niobium (T_c = 9.5K) have the highest transition temperatures of the elements. A strong electron–phonon interaction results in a high scattering level of electrons by the thermal vibrations and hence comparatively poor conductivity – but it does enhance the likelihood of superconductivity. By contrast the noble metals copper, silver and gold are good conductors because the scattering of electrons by phonons is weak – so weak an interaction that in fact it precludes them from being superconducting. A somewhat similar approach to constructing a theory based on electron–phonon interactions made independently in 1950 by Bardeen at the University of Urbana, Illinois in the U.S. also ran into difficulties. Fresh ideas were needed.

Working Towards a Successful Theory

The next crucial step towards an acceptable explanation of how superconductivity occurs at the microscopic level was made in 1956 by Leon Cooper – guided by Bardeen. The crucial realization is that superconductivity is associated with a bound pair of electrons, each having equal but opposite spin and angular momentum, travelling through the metal. Building on this idea, Bardeen, Cooper and Schrieffer, working at the University of Urbana, Illinois, produced a theory in which superconductivity is considered to arise from the presence of these "Cooper pairs". Of the three men, John Bardeen was by far the most senior and eminent. For much of his scientific career he had been intrigued by superconductivity. He recognized that a complete theory required the use of more sophisticated techniques such as quantum field theory, which was then just beginning to be introduced. To develop the required expertise, he had attracted Cooper, an expert in this area, to Urbana in 1955. Together with J. Robert Schrieffer, who had arrived at Illinois in 1953 as a graduate student from MIT, Bardeen and Cooper began a comprehensive assault on developing a microscopic theory of superconductivity. They were spurred on by the realization of competition from several other physicists studying the same problem, among them Richard Feynman, one of the most innovative and inspirational theoreticians of his generation.

Towards the end of 1956, success for the three seemed to be as far away as ever and Schrieffer confided in Bardeen that he was beginning to turn his efforts towards other more tractable problems. After all scientific progress is made by working on soluble problems! At the time, Bardeen was about to travel to Stockholm to receive the Nobel Prize for physics for the invention of the transistor. He received this award jointly with Walther Brattain and William Shockley for work that the three of them had carried out at the Bell Telephone Laboratories in Murray Hill, New Jersey in the late 1940s. Bardeen encouraged Schrieffer to continue with the superconductivity problem since he felt that they were close to success. The turning point came early in 1957 when they managed to deduce what is known as the correct ground state wave function for the superconducting electrons. This was followed by a frantic effort on the part of all three men as the details of the theory were worked out. After a preliminary note submitted to the journal *Physical Review* in February 1957, they worked on a much more substantial paper which appeared in the same journal in October of that year. This second paper, elegantly written and comprehensive, has become one of the classic papers of condensed matter physics, widely quoted and influential. The BCS theory accounted for many of the experimental observations, such as the existence of an energy gap $2\Delta(0)$ between the superconducting and normal states. A large number of experiments have confirmed this predicted value of the energy gap in the conventional superconductors. Recognition of the significance of their work came with the award of the 1972 Nobel Prize in Physics to Bardeen, Cooper and Schrieffer. For John Bardeen this was his second Nobel Prize in Physics, the only person ever to be so honored. For one person to develop the theory of both semiconductors and superconductors is a truly remarkable intellectual achievement and places Bardeen among the greatest physicists of the twentieth century.

In the next section the physical principles, which underlie the BCS theory are described in more detail. This requires a more theoretical argument; if this is not your scene at all, and you are happy with accepting the fact that exchanging phonons (heat) can hold a pair of electrons together then do not bother to read the accompanying boxes!

Physical Principles of the BCS Theory

One of the first steps to take when developing any theory of a physical phenomenon is to make an assessment of the energy involved. Superconductivity takes place at a lower temperature than normal state behavior; when a superconducting solid is heated above its critical temperature T_c, it goes into the normal state. Therefore, to drive a superconductor normal, energy is needed; this would be thermal energy, if the superconducting state is to be destroyed by increasing the temperature above the critical temperature T_c. However, it is also possible to drive a material normal by applying a magnetic field equal to a critical value. This magnetic behavior makes it easy to determine the energy difference between normal and superconducting states: all that is required is to measure the value of the critical magnetic field that destroys superconductivity. When this is done, it is found that the energy difference between the normal and the superconducting states is extremely small. For many pure metals, the critical magnetic flux density (B_c) at the 0K limit required to destroy super-conductivity is of the order of only 0.01 Tesla. This leads to an order of magnitude estimate of the condensation energy ($=\mu_0 B_c^2$) of only 10^{-8} eV per atom. To physicists struggling with the development of a theory of superconductivity, such a very small value of the superconducting energy presents a major obstacle: it is several orders of magnitude smaller than the energies involved in many processes always present in metals (for example, Coulombic interactions between the electrons lead to a com-paratively enormous correlation energy of the order of 1eV per atom). The small energy difference between the normal and superconducting states may also be compared with the energy of about 5eV for the conduction electrons in the normal metal. It is simply not possible to calculate the energy of the normal state electrons to the accuracy required to be able to separate off the tiny change due to the normal to superconducting transition. Any attempt at calculation of the condensation energy seems bound to fail because the much larger energy of other processes would be expected to mask that of the interaction responsible for the superconducting state. To avoid this dilemma, Bardeen, Cooper and Schrieffer assumed that the only important energy difference between the normal and the superconducting states arises from the interaction leading to superconductivity: they took the only reasonable theoretical approach of assuming that all interactions except the one causing superconductivity (i.e. Cooper pairing in the BCS theory) are unaltered at the normal to super-conducting state transition. They assumed that the only energy change involved when a material goes superconducting is that due to *the formation and interaction of the Cooper pairs.*

That electrons in a metal can pair at all is remarkable because they have the same charge and normally repel each other. So it is no surprise that the energy of electron pairing is extremely weak. In principle, only a small rise in temperature is enough to break a pair apart by thermal agitation and convert it back to two normal electrons. Nevertheless if the temperature is taken down to a sufficiently low value, the electrons do their best to get into the lowest possible energy states, so some pair off. The repulsion between electrons is overcome in two ways. First, some of the negative charge of an electron is blocked off or screened by the motions of other electrons. Second, an intermediary can bring the electrons together into pairs, which then behave more or less like extended particles. The first step in the formulation of a theory of superconductivity is to describe the nature of the interaction, which causes the pairs to form. A simple, commonly used analogy for such an interaction is given by two rugby football players, who can pair by passing the ball back and forth between them to avoid being tackled as they run up-field. The question is what corresponds to the ball in a superconductor? Answer: a phonon, the quantized packet of heat. You can find out more about phonons in Box 4.

Box 4

Phonons, the quantized packets of heat vibrations

Heat and sound are propagated in solids as thermal waves. Such lattice, or thermal, vibrations are waves propagated by displacement of the ion cores. The energy and momentum of these waves are quantized; thermal vibrations of frequency v_q may be treated as wave packets with energy hv_q. These quantized packets are called "phonons" by analogy to the "photons" of electromagnetic radiation. The word photon was devised from the Greek *photos* for light, phonon from that *phonos* for sound. Since a phonon has both direction and magnitude, it has to be described as a vector quantity $\mathbf{q}$, which is called the phonon wave-vector and has a value of $2\pi/\lambda$, λ being the wavelength of the associated thermal wave. This wave-particle duality of heat and sound arises as a result of the de Broglie hypothesis, which relates the momentum $\mathbf{p}$ ($=mv$) of a particle of mass m and velocity v and the wavelength λ, by:

$$p = h/\lambda,$$

where h is the Planck constant. Hence, since the value of $\mathbf{q}$ is $2\pi/\lambda$, the phonon momentum is $\hbar\mathbf{q}$, where the usual practice of writing $\hbar$ for $h/2\pi$ has been adopted. The energy of a phonon is much less than that of the conduction electrons in a metal, which are those electrons at the Fermi level and have the highest energy.

The discovery of the isotope effect suggested that interaction between electrons in states near the Fermi level and phonons is closely connected with the development of the superconducting state. In the case of an electron pair the "rugby football" being passed between two players and holding them together is a phonon. Electrons remain paired by exchanging phonons. The electron–phonon pairing mechanism, when embodied in the BCS theory, works extremely well for explaining superconductivity in conventional materials. The phonon acts as the "matchmaker" bringing the electrons together into pairs. The interaction between an electron and a lattice vibration can be treated as a collision between particles. Figure 6.1(a) illustrates this collision or scattering process, which is treated in more detail in Box 5.

Box 5

Interaction between electrons and phonons in Cooper pairs

On collision with a phonon, an electron of wavevector $\mathbf{k}$ absorbs the phonon and takes up its energy $h\nu_q$ (which is in general much less than that of the electron) and is scattered into a nearby state of wavevector $\mathbf{k}'$. Essentially, the electron has absorbed heat from the lattice and is now in a quantized state of different energy. The energy must be conserved in the process so that the new energy $E(\mathbf{k}')$ of the electron is the sum of its former energy $E(\mathbf{k})$ and that $h\nu_q$ of the absorbed phonon:

$$E(\mathbf{k}') = E(\mathbf{k}) + h\nu_q. \tag{1}$$

When the electron absorbs the phonon, it also takes up its momentum and changes its direction; this process is illustrated in Figure 6.1(a). Another basic law of the physical world has also to be obeyed: momentum must also be conserved:

$$\therefore \; \hbar\mathbf{k} + \hbar\mathbf{q} = \hbar\mathbf{k}' \; \text{or} \quad \mathbf{k} + \mathbf{q} = \mathbf{k}' \tag{2}$$

just as it would be for collision between two billiard balls. However, electrons are different from billiard balls: not only can an electron moving through a crystal lattice absorb phonons, it can also emit them. In this case, illustrated in Figure 6.1(b), the conservation of momentum leads to

$$\mathbf{k}' = \mathbf{k} - \mathbf{q}. \tag{3}$$

Particularly important in the BCS theory are so-called "virtual phonons". A solid can be thought of as teeming with virtual phonons, which exist only fleetingly. Indeed an electron moving through a lattice can be considered as continuously emitting and absorbing phonons: it is "clothed" with virtual phonons. Virtual states can be thought about in terms of the Heisenberg uncertainty principle in the form $\Delta E \Delta t \approx \hbar$. A phonon, which remains in a state for a time Δt, has an energy uncertainty ΔE. If the lifetime Δt of the phonon is very short, the energy uncertainty ΔE is very large and the phonon can transfer more energy than allowed by the law of conservation of energy. Over a period of time long compared with $\hbar/\Delta E$ energy must be conserved.

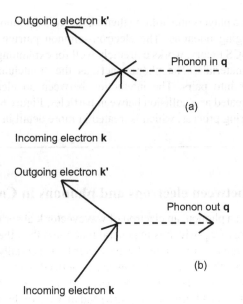

Figure 6.1   (a) Absorption of a phonon of wave-vector **q** by an electron in a state of wavevector **k**. The incoming phonon **q** is shown as the dotted arrow and the incoming (**k**) and outgoing (**k'**) electron as the filled arrows. (b) When an electron in a state **k** gives out a phonon of wave-vector **q**, it loses the energy of the phonon and goes into a new state **k'**.

Electron-virtual phonon processes play a central role in the development of the superconducting state. Cooper showed that electrons may be considered as being bound together in pairs by mutual exchange of virtual phonons. The process involved is illustrated in Figure 6.2. An electron in a state k_1 near the Fermi surface emits a virtual phonon **q** and scatters into a state k_1'. The law of conservation of momentum requires that for this process:

$$k_1' = k_1 - q. \tag{4}$$

Another electron in a state of k_2 absorbs the virtual phonon and is scattered to a state k_2' which is defined as:

$$k_2' = k_2 + q. \tag{5}$$

The two electrons, which exchange virtual phonons in this way, have interacted dynamically. Momentum must be conserved for the whole process; therefore from equations (4) and (5) above:

$$k_1 + k_2 = k_1' + k_2' = K. \tag{6}$$

Here **K** is the total momentum of the pair. In principle the interaction between the electrons may be either repulsive or attractive, the determining factor being

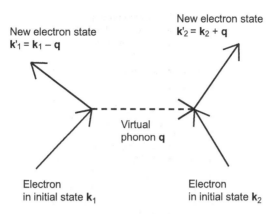

Figure 6.2 The process which binds two electrons into a Cooper pair. It is an interaction between the two electrons, which are in initial states with wave-vectors $\mathbf{k}_1$ and $\mathbf{k}_2$, by exchange of a virtual phonon of wavevector $\mathbf{q}$. The electrons go into new states $\mathbf{k}_1'$ and $\mathbf{k}_2'$.

the relative magnitudes of the phonon energy $h\nu_q$ and the energy difference between the initial and final states of the electrons. Cooper demonstrated that a weak attractive force could exist between pairs of electrons in a metal at low temperatures. For bonding between electron pairs to occur, the net attractive potential energy $(-V_{ph})$ arising from virtual phonon exchange must be larger than the Coulombic repulsive energy (V_{rep}) between the electrons. Therefore, using the convention that a negative potential energy gives rise to attractive forces, the energy balance being:

$$-V_{ph} + V_{rep} < 0.$$

A simple picture illustrates how an attractive force might arise between electrons in a lattice. As an electron moves through the lattice of positively charged ions, motion of the ions is disturbed in the near vicinity of the electron. The positive ions tend to crowd in on the electron: a screening cloud of positive charge forms around the electron. A second electron close by can be attracted into this region of higher positive charge density. The process in a two-dimensional square lattice is illustrated in Figure 6.3. If the ionic vibrations and the charge fluctuations produced by the first electron are in the correct phase, then the Coulombic repulsion between the two electrons is counteracted and the electrons are attracted into each other's screening clouds. The attractive energy between the electrons is increased when the electrons have opposite spin. By the process of exchanging phonons, the electrons in a Cooper pair experience mutual attraction at a distance.

The average maximum distance at which this phonon-coupled interaction takes place in the formation of a Cooper pair is called the *coherence length* ξ. In the early 1950s the Russian theorists Vitaly Ginzburg and Lev Landau produced an important phenomenological description of superconductivity, which had first introduced this concept of a coherence length. Their compatriot Lev Gorkov later showed that the

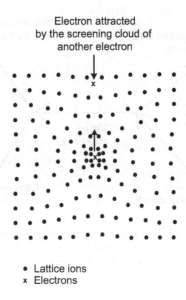

Electron attracted
by the screening cloud of
another electron

• Lattice ions
× Electrons

Figure 6.3 Attraction of an electron, shown by the cross (×), into the screening cloud of positive ions pulled in towards another electron (×) at the center of the picture. The sketch is very diagrammatic. The process of interaction is dynamic and both electrons distort the lattice.

Ginzburg–Landau theory can be derived from the BCS theory: both give equivalent results close to the superconducting critical temperature. The coherence length is fundamental to superconductivity and emerges as a natural consequence of the BCS theory.

One way of estimating a value for the coherence length is to apply a nearly critical magnetic field to a superconducting sample. Then parts of the sample near the extremities go into an intermediate state, a laminar structure composed of both normal and superconducting regions. The boundary between each normal and superconducting region is not sharp but has a finite width; physical properties also vary through the boundary. Careful examination of the boundary shows that it results from the long range of influence of the superconducting electrons over a macroscopic distance of about 10^{-4} cm., which is the coherence length.

Long-range order or coherence takes place between electrons in superconductors and is a measure of the sphere of influence of a Cooper pair. Coherence suggests that the waves associated with the pairs are macroscopic in extent and overlap considerably with each other.

This coherence results in a superconductor behaving rather as if it is a "giant molecule" i.e. an "enormous quantum state". In the early nineteenth century, when Ampère had proposed that magnetism can be understood in terms of electric currents flowing in individual atoms or molecules, it was objected that no currents were known to flow without dissipation. He has long since been vindicated by quantum theory, which gives rise to stationary states in which net current flows with no resistance. A superconductor is a dramatic macroscopic manifestation of a quantum mechanical state, which behaves like a giant molecule with no obstruction for electron flow: there is no resistance.

The Superconducting Energy Gap: a Fundamental Difference Between the Arrangements of the Electron States in Superconducting and Normal Metals

A crucial feature of the superconducting state is the existence of an energy gap at the Fermi level region of the excitation spectrum of superconductors. Electrons are not allowed to possess energies within this forbidden range of energy. The Cooper pair states exist just below the energy gap. The energy gap corresponds to the energy difference between the electrons in the superconducting and normal states. The confirmation of this long suspected feature of the arrangement of the states available for electrons in superconductors was a decisive step in the development of an understanding of superconductivity. This energy gap arises as a result of the interaction of the Cooper pairs to form a coherent state in which the superconducting electrons have a lower energy than they would have in the normal state. More formally, a central prediction of the BCS theory is that the Cooper pairs form a condensed state whose lowest quantum state is stable below an energy gap of value 2Δ, which separates the superconducting states from the normal ones. An important test of the BCS theory was to measure this gap and compare the value obtained with that predicted.

At an early stage it was noticed that a superconductor looks the same as the normal metal: there is no change in its appearance, if a metal is cooled below the critical temperature. This means that the reflection and absorption of visible radiation by a superconductor are the same as those in the normal state. However by contrast, a superconductor shows great differences in its response towards flow of a.c. and d.c. currents from those found in the normal state. Unlike normal metals, superconductors exhibit no resistance to direct current flow; but towards an alternating current they do show some resistance and this increases as the frequency goes up. If the alternating frequency lies in the infra-red region above about 10^{13} cycles/sec or beyond into the visible light range, superconductors behave in a similar manner to normal metals and absorb the radiation. That is why they look the same as normal metals in visible light.

This difference between the behavior of a superconductor towards high and low frequency provides evidence for the existence of the energy gap and suggests one way of measuring it. In normal metals, when photons of electromagnetic radiation are absorbed, electrons are excited into stationary states of higher energy. A fundamental property of superconductors is that they can absorb electromagnetic radiation only above a threshold frequency. Behavior of this type is characteristic of materials with a gap containing no allowed energy states in the energy spectrum. Electrons in states just below such an energy gap (usually said to be of value 2Δ) cannot be excited across the gap unless they absorb a photon of sufficient high energy to enable them to bridge the gap completely. The threshold frequency ν_g for absorption of radiation

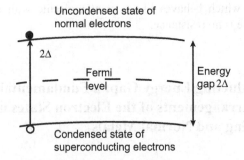

Figure 6.4 In a superconductor the energy gap is centred at the Fermi level. This diagram is an expanded small part close to the surface of the sphere shown in Figure 5.7.

is given by $2\Delta/\hbar$. Absorption of radiation beyond the threshold frequency v_g by a superconductor occurs because pairs of electrons in the condensed superconducting state are excited by the radiation across the energy gap into states in which electrons exhibit normal behavior. From infrared absorption experiments, among others, the measured gap (2Δ) is found to be in agreement with the BCS predicted value of about $3.5k_BT_c$, k_B being Boltzmann's constant and T_c the critical temperature. The energy gap is centered at the Fermi level, as illustrated in Figure 6.4. The width of the energy gap is such that photons with a high enough energy to enable Cooper pairs to be split and surmount this threshold energy ($\hbar v_g$) lie in the short microwave or the long infrared region, until recently a range of the electromagnetic spectrum not readily accessible to experiment (see Chapter 9). Therefore, the gap is not that easy to observe and so its discovery was long delayed. Recognition of the existence of the gap gave a much clearer picture of the structure of the energy states in super-conductors and an important indication of the type of theory necessary to explain superconductivity. In summary, in a superconductor the electron pairs form a con-densed state below an energy gap that separates the superconducting states from those available for normal electrons.

The BCS Model of a Conventional Superconductor

Cooper had shown that there can be a small net attractive force between pairs of electrons; hence pairs are able to exist at low temperatures. Below the critical tem-perature, pairing of the electrons close to the Fermi surface is the more stable con-figuration in the superconducting state and reduces the total energy of the system. This is why pairing takes place. An electron pair does not behave like a point particle but instead its influence extends over a distance of about 10^{-4} cm. in agreement with the experimental measurements of the coherence length. In consequence the volume of a Cooper pair is about 10^{-12}cm^3. But there are about one million other Cooper pairs in this region: the spheres of influence of the pairs overlap extensively. It is no longer possible to talk about isolated pairs because the electrons continually exchange partners with each other; that is the same as saying that the pairs interact with each other. Overlap between the waves of the two electrons in a pair and then in turn between the waves associated with the pairs results in the coherence and produces the condensed state in a superconductor. The electron pairs collect into what may be likened to a macromolecule extending throughout the metal and capable of motion as a whole.

The BCS theory is based on this interaction between pairs of electrons to form a giant quantum state. A superconductor can be visualized as a complex square dance of Cooper pairs which are all moving in time with each other and exchanging partners continuously. This "*dance to the music of time*" comprises the condensed state that has more order and is lower in energy than that of the electrons in a normal metal. BCS propose, as the criterion for the formation of the superconducting state, that Cooper pairs are produced at low temperatures and that this is the only interaction in a superconductor that results in an energy different from that of the normal state. To simplify the problem, BCS calculate the superconducting properties for the simple model of a metal having a spherical Fermi surface, which has been described in Chapter 5 and illustrated in Figure 5.7. They make the further simplification that only those electrons near the Fermi surface need be considered in the formation of the condensed superconducting state. If the average phonon frequency in the metal is v_g, the electrons, which can be bound together into Cooper pairs by exchange of phonons, are those within an energy $\hbar v_g$ of each other. Electrons within this small energy range near the Fermi surface are bound together in pairs while all the others outside this thin shell remain unpaired. This abrupt, somewhat arbitrary, cut-off usually gives satisfactory results. In fact, subsequent work indicates that the results of the BCS theory are not particularly sensitive to the form of the cut-off.

Just as all ideal gases obey Boyle's law, conventional superconductors comply with the BCS theory and behave in the same general fashion as each other.

In the BCS model the coupled pairs of electrons have opposite spin and equal and opposite momentum (see Box 6) and are condensed into a giant state of long-range order extending through the metal – such an extraordinary feature of the super-conducting state that it needs to be considered further separately (Chapter 7).

Since all the pairs are in harmony with each other, the whole system of cor-related electrons resists rupture of any single pair. Therefore, inherent to the system is the property that a finite energy is necessary to break up only one pair. In a normal metal, electrons at the Fermi surface can be excited by what is, to all intents and purposes, an infinitesimally small energy, whereas in a superconductor pair cor-relation produces a small but finite energy gap, whose value $2\Delta(0)$ at the absolute zero is given by a famous BCS formula relating the gap to the critical temperature T_c

$$2\Delta(0) = 3.5k_B T_c.$$

The 2 comes about in the left hand side of this equation because a pair of elect-rons has to be broken up for energy to be absorbed. Excited, single-particle, "normal" states are separated from the correlated pair states by this energy gap. A single electron is a fermion. But a bound Cooper pair acts as a boson; this is because

---

Box 6

Electrons in a Cooper pair have opposite momentum and spins

When the bonding between pairs is as strong as possible, the system is at equilibrium because the energy is at a minimum. Interaction between the pairs is
strongest when the number of transitions of electrons from pair to pair is as large
as possible. This occurs when the total momentum of each pair is the same as
that of any other pair. The condition satisfying this is that the total momentum $\mathbf{P}$
is zero. Now by the de Broglie hypothesis $\mathbf{P}$ is equal to $\hbar\mathbf{K}$, so that the total wave
vector $\mathbf{K}$ (see Box 5) of each electron pair is also zero. In this situation, the
electrons in each pair must have equal and opposite momentum. Then

$$\hbar\mathbf{k} + \hbar(-\mathbf{k}) = \hbar\mathbf{K} = 0 \quad \text{or} \quad \mathbf{k} + (-\mathbf{k}) = 0.$$

A further prerequisite for minimum energy is that the electrons in each pair
have opposite spin. The net spin on a pair is zero. This means that the pairs are
bosons. Now bosons do not obey the Pauli exclusion principle so that they can
all occupy the same state (see Chapter 5). In a superconductor all the Cooper pairs
are condensed into the same state. Hence it is possible to depict the arrangement
of the electron states by the simple energy level diagram shown in Figure 6.5.

---

if both electrons in a pair are changed, the sign of the wave function is altered twice
and thus is unchanged (theoreticians say that it is invariant under this transformation).
Since the pairs are bosons, all of them are contained in the same state of lowest
energy. The energy levels into which electrons can go for a superconductor can be

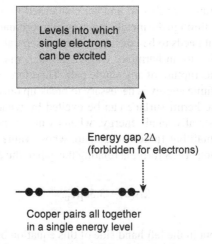

Figure 6.5   The arrangement of the energy levels in a superconductor. All the Cooper
pairs collect into a single level, which is separated by the energy gap from the higher states
into which single electrons can be excited.

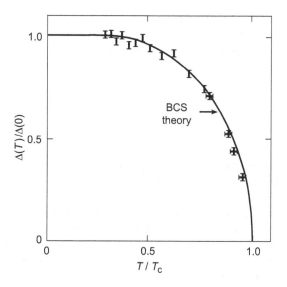

Figure 6.6 The way in which the superconducting energy gap $2\Delta(T)$ varies with temperature T. The energy gap $2\Delta(T)$ at a temperature T has been divided by that $2\Delta(0)$ at the absolute zero. This reduced energy gap $\Delta(T)/\Delta(0)$ has then been plotted as a function of the reduced temperature T/T_c. The full curve shows the BCS prediction of the temperature dependence of the energy gap. The experimental points are those found for an indium–bismuth alloy and are in good agreement with the theoretical prediction.

shown in a simple way (Figure 6.5). The Cooper pairs all exist in one level separated by the energy gap 2Δ from a higher energy band of single levels into which normal electrons (from split pairs) can be excited.

That the BCS theory should result in an energy gap and predict its magnitude was one of its major triumphs. For conventional superconductors experimental measurements of the energy gap are in good agreement with this BCS theory prediction. For example for aluminium T_c is 1.14K and the measured gap extrapolated to the absolute zero of temperature is $3.3k_BT_c$ ($\equiv 3.4 \times 10^{-4}$eV) while the BCS theory predicts $3.5k_BT_c$.

At any finite temperature there are always a few electrons which have been excited thermally across the gap. This reduces the number of electron pairs and the correlation energy becomes correspondingly less. Therefore, as the temperature T is increased, the energy gap $2\Delta(T)$ becomes smaller, as illustrated in Figure 6.6. At the critical temperature the energy gap vanishes, there are no pairs and the normal state is assumed.

The BCS model bears a strong resemblance to the earlier two-fluid model pioneered by Gorter and Casimir (see Chapter 2). In a superconductor at any finite temperature below the critical temperature T_c, there are two different kinds of electron states. Occupied excited states above the gap contain single electrons, while in the condensed, superconducting state below the gap the electrons are paired and the pairs are correlated. When pairs are present, they short-circuit the normal electrons: the pairs carry the "supercurrent". Now that we have acquired some of the basic ideas about the nature of superconductors, we can have a look at the mechanism by which the electrons in pair states can carry a "supercurrent" without resistance – that sensational discovery made by Kamerlingh Onnes.

Zero Resistance and Persistence of Current Flow in a Superconductor

A successful microscopic theory of superconductors must provide an adequate description of the mechanism of resistanceless flow of persistent current. The BCS model does this.

Current flow in a superconductor, which is described in more detail in Box 7, resembles that in normal metals, save for one fundamental difference: individual supercurrent-carrying electrons cannot be scattered. In the superconducting state all the electron pairs have a common momentum and there is long-range correlation of momentum. The usual situation is that resistance to current flow can only occur when scattering processes transfer electrons into empty, lower energy states with momentum in the opposite direction to the electron current. Although this process occurs in normal metals (Chapter 5 (Box 3, illustrated in Figure 5.10), it can not take place in superconductors. Figure 6.7(a) shows schematically the occupation of states in a superconductor which is not carrying current; states with opposite momentum at the Fermi level are bound into Cooper pairs. When a current is passed, there is increased momentum in the direction of the flow of Cooper pairs, as shown in Figure 6.7(b). If the electrons in the highest energy states on the right hand side of Figure 6.7(b) could be scattered into the empty states on the left hand side, this would lead to a decrease in momentum along the direction of electron current flow: there would be electrical

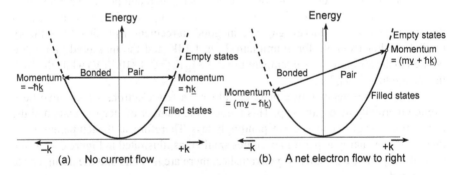

(a)   No current flow   (b)   A net electron flow to right

Figure 6.7   The effect of an electric field on the occupation of the available **k**-states in a one-dimensional superconductor. The full lines drawn on the lower energy states in the parabolic bands show states filled with electrons, while the dotted lines at higher energies represent empty states. The double headed arrows connect electron pairs with opposite momenta. In case (a) there is no applied field, while there is an applied field in case (b). For scattering to take place, electron pairs must be disrupted to allow electrons in the higher energy states on the right hand side to go into the empty, available, lower energy states on the left hand side. This cannot happen.

resistance. However, such scattering processes would necessitate the splitting up of electron pairs and their removal from the condensed state. It is just this process that the correlated electron pair system resists most strongly. For a pair to be broken both electrons must obtain sufficient energy to be excited across the energy gap and disturb the entire correlated system. In a superconductor the energy available from any single scattering processes by phonons (or other mechanisms) is not enough to do this. Scattering is suppressed: there is no resistance. Since the correlated pair system opposes change, current flow by the Cooper pairs is persistent.

To describe persistent current flow, Bardeen used a colorful analogy, referring to a closely packed crowd that has invaded a football field. The Cooper pairs can correspond to couples in the throng, who are desperately trying to remain together. Such a crowd is hard to stop – once set in motion – since stopping one person in the group, requires stopping many others. The crowd members will flow around obstacles, such as goalposts, with little disruption – suffering no resistance.

Box 7

Zero resistance and persistence of current flow in a superconductor

When a superconductor is not carrying a current, there is no net pair drift velocity because as many electrons go one way as the other: a pair has zero net momentum (Figure 6.7(a)). One electron in a pair has momentum $\hbar k$ and the other $-\hbar k$. When a current is flowing, the net drift velocity is v and net momentum mv along the direction of electron flow (Figure 6.7(b)). The energy of the current-carrying state is higher than that of the ground state. Coupling by virtual phonons now takes place between electrons with momenta $(mv + \hbar k)$ and $(mv - \hbar k)$. The total wave vectors of every pair (of total mass $2m$) are all equal (as they were before) *but now are no longer zero.* If

$$v = \hbar P/2m \quad \text{or} \quad mv = \hbar P/2$$

so that the supercurrent-carrying states are translated in **k**-space by the wave-vector $\hbar P/2$, then the pairs have wave vectors of $(\mathbf{k} + \mathbf{P}/2)$ and $(-\mathbf{k} + \mathbf{P}/2)$. Therefore, the total wave vector of each of the pairs is:

$$(\mathbf{k} + \mathbf{P}/2) + (-\mathbf{k} + \mathbf{P}/2) = \mathbf{P}$$

and the pair momentum is $\hbar \mathbf{P}$. Scattering requires that a pair of electrons is broken up and this can only happen if a minimum energy 2Δ is supplied from somewhere to take both electrons across the gap. At low current densities, this amount of energy can not be given to the electron pairs. Scattering events which change the total pair momentum are inhibited; there is no resistance.

Another way of understanding the persistence of current flow is as follows: to take a pair of electrons away is very difficult because of the tendency of bosons to keep together in the same state. Once a current is started it just keeps going forever.

In general, the way in which superconductors behave arises because the electron pairs are bound together in a single energy level and resist removal from it. The BCS theory does account for persistence of a supercurrent. Not only can most of the known facts about superconductivity in conventional materials be explained but new properties are also predicted.

Experimental Examination of the Electron Pair Theory

An energy gap in the elementary excitation spectrum of the electrons in super-conductors was postulated several years before the development of the electron pair theory. Nevertheless, the results obtained by BCS, which predicted the magnitude and temperature dependence of the energy gap are an outstanding theoretical achievement. Examination of these predictions is the most obvious way to verify the theory experimentally. Spectroscopic measurements of microwave (Chapter 9) and infrared absorption give direct evidence for the gap and its magnitude; at low frequencies there is no absorption but at a frequency v_g such that hv_g equals the energy gap at the temperature of measurement there is a sharp onset of absorption of the radiation: an absorption edge is observed. Thermal properties, such as specific heat and thermal conductivity, related to the energy required to excite electrons across the gap, also provide valuable information about the energy gap.

Perhaps the most striking confirmation of the energy gap comes from *electron tunnelling* experiments. Tunnelling refers to the fact that an electron wave can penetrate a thin insulating barrier, a process that would be forbidden under the laws of classical physics. If a ball is thrown against a wall, it bounces back, but an electron has a probability that it can tunnel through a forbidden region. A fraction of the electrons moving with high velocities in a metal will penetrate a barrier by tunnelling, producing a weak tunnel current on the other side of the barrier. In the late nineteen twenties some phenomena in solids were explained by tunnelling but progress in using tunnelling was slow until 1958, when a young Japanese physicist Leo Esaki at Sony Corporation pioneered the initial experiments that established the existence of the effect in semiconductors. In 1960, an engineer Ivar Giaever, working on electronic devices made by thin film technology at the General Electric Research Laboratory in Schenectady, New York, conjectured that tunnelling might also be used to great effect in the study of superconductors. In particular, he suggested that the energy gap could be measured from the current–voltage relation obtained by tunnelling electrons through a thin sandwich of evaporated metal films insulated by an oxide film. Experiments showed that his conjecture was correct and tunnelling became the dominant method of determining the energy gap in superconductors. Later Brian Josephson predicted tunnelling of Cooper pairs through a thin insulating barrier (this will be discussed in Chapter 7). In 1973 Esaki, Giaever and Josephson were awarded the Nobel Prize in Physics for their discoveries of electron tunnelling phenomena in solids.

Ivar Giaever has told of his trials and tribulations in an amusing way in his Nobel Prize lecture: neither he nor his colleague John Fisher

> "had much background in experimental physics, none to be exact, and we made several false starts. To be able to measure a tunnelling current the two metals must be spaced no more than about 100Å apart, and we decided early in the game not to attempt to use air or vacuum between the two metals because of problems with vibration. After all, we both had training in mechanical engineering! We tried instead to keep the two metals apart by using a variety of thin insulators made from Langmuir films and from Formvar. Invariably, these films had pinholes and the mercury counter electrode, which we used, would short the films. Thus we spent some time measuring very interesting but always non-reproducible current–voltage characteristics, which we referred to as miracles since each occurred only once. After a few months we hit on the correct idea: to use evaporated metal films and to separate them by a naturally grown oxide layer."

To prepare a tunnel junction without pinholes, these early workers first evaporated a strip of aluminium onto a glass slide. This film was removed from the vacuum system and heated to oxidize the surface rapidly. Several cross strips of aluminium (Al) were then deposited over the first film making several junctions at the same time. In this way capacitors, plate electrodes separated by a thin oxide film, were made with superconducting film. Tunnel devices now have important uses and further details of their manufacture are detailed more appropriately in Chapter 9. To obtain the current–voltage characteristic of a tunnel junction, a voltage was applied across it in the circuit shown in Figure 6.8 and the current flow measured. By April 1959, successful tunnelling experiments had been carried out that gave reasonably reproducible current–voltage characteristics. A typical current–voltage characteristic for tunnelling between two superconductors is shown in Figure 6.9; only a minute current can flow across the junction until the voltage applied is sufficient to excite Cooper pairs across the superconducting energy gap.

To explain how the energy gap can be determined, it is instructive to discuss an experiment on a capacitor which has one plate made from a superconductor while the other is a normal metal. When the capacitor plates are less than 100Å apart, a quantum mechanical tunnelling current can flow across the device. Conduction electrons in the metal plates behave as running waves, which are reflected back into the metal at the surfaces; however, there is a finite probability that an electron, on one of many "trial runs" at the surface, may tunnel through the thin layer of insulator separating the plates. The way in which the energy levels are arranged when no voltage is applied to a capacitor which has one superconducting and one normal metal plate is shown in Figure 6.10(a). As shown by the energy level diagram for a superconductor (Figure 6.5), the Cooper pairs are all contained in one level separated from the excited states by the energy gap (value 2Δ). When there is no voltage applied across the capacitor, the level containing the Cooper pairs in the superconductor is

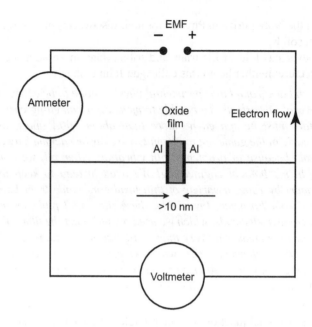

Figure 6.8 A circuit diagram for making a tunnel experiment. The device is like a capacitor whose electrodes are the superconductor under investigation (in this case aluminium (Al)) separated by a very thin, insulating aluminium oxide film. To measure the current−voltage characteristics of the device, a variable voltage (labelled EMF) is applied across it and measured using the voltmeter; the resultant current is measured with the ammeter.

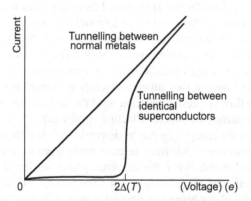

Figure 6.9 The experimental current−voltage characteristic observed for tunnelling between identical superconductors. There is a sharp increase in the tunnelling current when the applied voltage is equivalent to the energy gap at the temperature of measurement. For comparison tunnelling between two normal metals is also shown.

lined up with the Fermi level in the normal metal on the other side of the insulating barrier. Electrons can only tunnel through the insulating layer, if there are empty states for them to go into; no current flows. However, when a positive voltage is

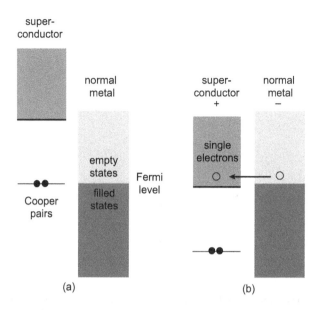

Figure 6.10   Energy level diagram for tunnelling across a device made of a superconductor separated from a normal metal by a thin insulating film. (a) No applied voltage. There are only a few excited electrons above the superconducting energy gap. The Cooper pair level is at the same energy as that of the Fermi level. (b) When the applied voltage is large enough for the states containing single electrons which face empty states above the Fermi level of the normal metal on the other side of the junction, a substantial current can flow because electrons can tunnel across into the empty states then available for them.

applied to the superconductor, the Cooper pair state is lowered (Figure 6.10(b)). Tunnelling still does not take place until the voltage becomes large enough for the edge to be pushed to the same level as the Fermi energy in the normal metal. Electrons at the Fermi level in the normal metal can now tunnel from the negative plate across the insulator into the empty excited states in the superconductor that forms the positive plate. A current flows.

Hence to measure the energy gap, a positive voltage V is applied to the superconducting plate. This lowers the energy levels of the superconductor relative to those in the normal plate. For a small voltage no tunneling occurs. However, when the applied voltage V is raised to a value $V_{critical}$ equal to Δ/e, the Fermi level in the normal metal is lifted up to the lowest empty excited states in the superconductor. So at this applied voltage, which is a direct measure of one half Δ of the gap, there is a sharp increase in the current flowing across the capacitor. The current–voltage characteristic of the device is illustrated in Figure 6.11. In effect, tunneling experiments allow direct measurements of the gap with a voltmeter. The voltage is small, only of the order of a millivolt.

Giaever also observed a characteristic fine structure in the tunnel current, which depends on the coupling of the electrons to lattice vibrations. From these beginnings tunnelling has now developed into a spectroscopy of high accuracy to study in detail the properties of superconductors. The experiments have confirmed in a striking way

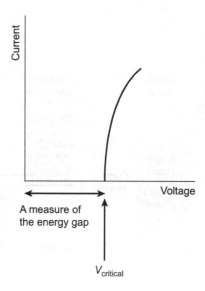

Figure 6.11 The experimental current–voltage characteristic observed for tunnelling across a capacitor made with one plate of normal metal and the other plate of super-conductor. At a voltage less than that required to break Cooper pairs and excite normal electrons above energy gap no tunnelling current flows. When the voltage $V_{critical}$ is large enough to break pairs, there is a sharp increase in the tunnelling current. The critical voltage $V_{critical}$ can be used to measure the energy gap ($V_{critical} = \Delta/e$) at the temperature of measurement. Here Δ is half the gap and e is the electronic charge.

the validity of the BCS theory: extrapolated values of the limiting energy gap at 0K are between $3kT_c$ and $4.5kT_c$ for conventional superconductors. The temperature dependence of the gap usually follows quite closely the BCS predicted curve shown by the solid line in Figure 6.6.

Several other techniques have been used to determine the energy gap in super-conductors. One of these is to measure the absorption of monochromatic, ultrasonic waves in the frequency range 10MHz to 1000MHz in which the ultrasound wave energy is small compared with the energy gap. Normal electrons scatter the ultra-sound waves, attenuating them. So as the temperature is lowered through the critical temperature T_c, reducing the number of normal electrons, the attenuation of ultra-sound waves decreases sharply. Determinations of the energy gap from ultrasonic attenuation measurements are in reasonable accord with BCS predictions. In 1964, while at the University of California, Riverside, George Saunders made ultrasonic attenuation measurements on metal single crystals, and discovered that the energy gap is anisotropic: it depends upon crystallographic direction. Typical results of the anisotropy of the energy gap are detailed in Table 6.1. This directional effect is a result of the Fermi surface also being anisotropic. It shows one limitation of the BCS theory, which, as we have seen, is based upon a spherical model of the Fermi surface and therefore cannot predict anisotropy of the energy gap.

There are also problems in the application of the BCS theory to alloys and other more complex superconductors. Phil Anderson suggests that in alloys strong scattering results in a more nearly constant interaction than that for anisotropic, pure metal superconductors. In these "dirty" superconductors the energy gap should be isotropic and the BCS theory should be obeyed closely. Measurements made by George Saunders of ultrasonic attenuation in the intermetallic, disordered alloy of composition equivalent to In_2Bi confirm this prediction. The energy gap at 0K is $(3.4\pm0.2)k_BT_c$ and, as shown in Figure 6.6, the temperature dependence of the energy gap is in reasonable agreement with that predicted by BCS. Other experimental measurements also suggest that the BCS theory is applicable to alloys. Only a few of the experimental methods used to examine the BCS theory have been surveyed here. There are numerous other techniques available. The results confirm that the generalized BCS model of superconductivity is essentially correct for many alloy superconductors.

Interesting discoveries in the field of conventional superconductors are still being made. As recently as January 2001, Professor Jun Akamitsu of the Aoyama Gakuin University in Tokyo announced at a symposium on "Transition Metal Oxides" in Sendai, Japan, that magnesium boride (MgB_2) is a binary intermetallic superconductor with a critical temperature T_c of 39K – the highest yet found for a conventional material. This discovery caused an immediate flurry of excitement as scientists worldwide attempted to verify and extend the observation. Just as the aftermath of the discovery of high temperature superconductors by Bednorz and Müller meant that many scientists communicated with each other via faxes, pre-prints and telephone conversations, so the more recent discovery led many scientists to first announce their results on the Internet; a trend which it seems is set to continue. By examining a sample containing the less abundant boron isotope (^{10}B), a group led by Paul Canfield at the Ames Laboratory in Iowa State University has already shown that MgB_2 has an isotope effect, which is consistent with the material being a phonon-mediated BCS superconductor, although such a high transition temperature might have implied an exotic coupling mechanism. Strong bonding with an ionic component and a considerable electronic density of states produce strong electron–phonon coupling, and in turn the high T_c. Other experiments such as electron tunnelling are also consistent with a BCS mechanism.

Table 6.1 The limiting energy gap at 0K in different directions in thallium (Tl) and tin (Sn) as determined from acoustic attenuation measurements. The directions in column two are given in Miller indices which are defined in standard texts on crystallography.

	Direction of propagation of ultrasonic waves	Energy gap $(2\Delta(0)/k_BT_c)$
Tl	$[10\bar{1}0]$	4.1
	$[1\bar{2}10]$	4.0
	$[0002]$	3.8
Sn	$[001]$	3.2
	$[110]$	4.3
	$[010]$	3.5

It is possible to purchase MgB_2 directly from chemical suppliers and many of the first superconducting samples were obtained in that way. The recent discovery does pose the question as to why the superconducting properties of MgB_2 were not discovered years ago. In Chapter 2, the systematic search for new superconductors by John Hulm, Berndt Matthias and others was discussed. They had great success with intermetallic compounds based on transition metals but failed to find super-conductivity in any transition metal diborides that they examined. The recent discovery undoubtedly means that there will be renewed interest in other binary and ternary intermetallic compounds. Finally, an interesting ramification is that MgB_2 can be thought of as an analogue of the predicted metallic hydrogen superconductor, which many believe could have an extremely high critical temperature T_c, and which may exist in cold stars.

Summary

Within the limits imposed by the relative simplicity of the model, the BCS theory provides an acceptable explanation of the phenomenon of superconductivity in the conventional materials. The BCS model extends the concept of a "two-fluid" super-conductor. At temperatures below the critical temperature T_c there are both normal and superconducting electrons. Intrinsic to the superconductor is an energy gap between the two types of particle states. The superelectrons consist of pairs of electrons coupled by phonons. Overlap between pair waves gives rise to a condensed state of long-range order capable of sustaining persistent currents; in the super-conducting state, quantum effects are acting on a macroscopic scale. Experimental results on pure metal superconductors are in reasonable agreement with the theory. Certainly, measured values of the limiting energy gap at 0K ranging from $3.2k_BT_c$ to $4.6k_BT_c$ are in accordance with the prediction of $3.5k_BT_c$, and the measured tem-perature dependence of the gap is in general in keeping with the theory. Real metals are more complex than the idealized BCS model. The theory is framed to deal with the general cooperative nature characteristic of all superconductors. Superconduct-ors, however complicated their energy surfaces may be in reality, are treated within the context of the same model: the BCS model is really a *law of corresponding states*.

All conventional superconductors show some departures from the BCS super-conductor, but deviations are surprisingly small. Recourse to a stronger electron–phonon coupling interaction than that used in the initial theory can resolve many of the difficulties that do arise. For instance, in the strong-coupling limit the predicted energy gap of $4.0k_BT_c$ at 0K, as against the $3.5k_BT_c$, of the weak coupling theory of BCS, accords with the experimental data for mercury and lead. Thus a more realistic choice for the coupling interaction can allow for some variation in behavior from metal to metal. Agreement between experiment and theory is then much closer. The electron-pair hypothesis occasions a point of departure rather than a conclusion to the subject. Not only are known facts explained but also new phenomena are predicted.

One requirement of a theory of superconductivity is that it should predict which materials may be superconducting. In this BCS is somewhat reticent. Nevertheless, the theory does suggest that a strong interaction between the lattice and electrons is conducive to the formation of the condensed state of Cooper pairs. A strong electron–phonon interaction inhibits normal state conduction because the electrons are more strongly scattered: metals such as tin, lead, thallium and mercury, which are relatively poor conductors in the normal state, tend to be superconductors, while the best conductors of electricity, the noble metals, in which lattice scattering is weak, do not become superconductors.

7 The Giant Quantum State and Josephson Effects

One of the most fascinating and fundamentally important properties of superconductivity is its quantum behavior over large distances. Usually quantum mechanical effects are only important at low temperatures and over distances on the atomic scale, that is about 10^{-9} meters. Superconductors are an exception to this rule. As long ago as the late 1940s, Fritz London, with a great leap of the imagination suggested that for superconductors, the wave–particle duality should be able to be seen in vastly larger objects, even to a mile long superconducting loop.

As for all matter, the de Broglie hypothesis applies to Cooper pairs of electrons: there is a wave associated with them. The de Broglie wave of a Cooper pair extends over a distance of about 10^{-6} meters – some thousand times longer than the spacing between atoms in a solid. This size scale of a Cooper pair defines a coherence distance between the individual electrons forming the pairs. The essence of superconductivity is coherence between the de Broglie waves of *all* the Cooper pairs: this phase coherence extends over the whole of a superconducting body even though it may be enormous. Phase coherence corresponds to BCS telling us that *all the Cooper pairs behave in exactly the same way,* not only as regards their internal structure but also as regards their motion as a whole: they all move in time with each other. In a Cooper pair, the electrons are bound together to form an entity rather akin to a "two electron molecule". Each pair can be thought of as a wave (Figure 7.1) travelling unscattered throughout the whole volume of the superconducting metal. Each electron finds itself preferentially near another electron in a Cooper pair, which acts over such a large volume that within it there are millions of other electrons each forming their own pairs. As a result, the waves, now relating to the whole collection of pairs, overlap in a coherent manner (Figure 7.2): in addition to having the same wavelength, the pair waves are all in step: they have the same phase in time.

When there is a superconducting current flowing in a metal, all the pairs have the same momentum; the corresponding waves all have the same wavelength and all travel at the same speed. These waves (Figure 7.2) superpose on each other to form a synchronous, co-operative wave with that same wavelength. Whatever the size of the superconductor, all the electron pairs act together in unison as a wave that shows the extraordinary feature of remaining coherent over an indefinitely long distance spanning the entire superconductor. This property, known as phase coherence, is central to superconductivity and has profound consequences: indeed it can be thought of as being responsible for the curious properties of superconductors. It leads to the existence of the superconducting energy gap and is the source of the long-range order

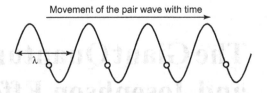

Figure 7.1 Travelling wave. A Cooper pair can be represented by a *travelling wave*. A sine wave of wavelength λ is used here as a simple way to enable "visualization" of the Cooper pair wave. If **P** is the momentum of the pair, the *de Broglie wavelength* λ of this travelling wave is $h/\mathbf{P}$. The open circles represent points of equal *phase;* this time phase is associated with the energy of the pair. Physicists use the *wave function* ψ as a mathematical tool to represent particles in quantum systems. Like any wave, this function has both amplitude and phase. $|\psi^2|$ gives the probability for a particle to be in a particular place at a particular time.

of the superconducting electrons. It is the reason why superconductors exhibit quantum effects over large distances. This coherent wave, which is identical for all the pairs throughout the superconductor, can undergo interference and diffraction effects, analogous to those observed for light waves, that are manifest in the macroscopic quantum interference effects, observed in SQUIDS (see Chapter 9) and have useful applications.

So each superconducting pair is characterized by a wave with an amplitude and a phase. The superconducting ground state is a coherent superposition of pairs – all having the same phase. Let us consider what happens if a current of electrons is set up round a ring. Motion around a ring made of a metal in the normal state causes electrons to accelerate centripetally (in an analogous manner to the moon travelling in orbit around the earth) and they continuously emit electromagnetic radiation and lose energy. By contrast, in a superconducting ring a supercurrent persists and does not lose energy by radiating electromagnetic waves. Stationary states of this kind, which do not alter with time, are governed by quantum conditions. These require the quantization of the energy of the superconducting current. This situation is analogous to the quantization of the energy levels for an electron in orbit around the proton in a hydrogen atom: in the Bohr model (see Figure 5.1) an electron remains indefinitely in its orbit with an unchanged energy and does not radiate electromagnetic waves. To increase its energy, the electron has to jump into another quantum state having a higher fixed energy. Similarly a supercurrent in a ring is quantized and can only be increased by a jump up into a state of higher fixed energy and current. One consequence of this quantization of the current round a ring is that the magnetic flux threading through the ring is quantized and we now consider the ramifications of this remarkable feature.

Flux Quantization

Flux quantization is another quantum effect of superconductors that was predicted by Fritz London. It was thought to be so bizarre that nobody paid much attention to

Cooper pair wave 1

Cooper pair wave 2

Cooper pair wave 3

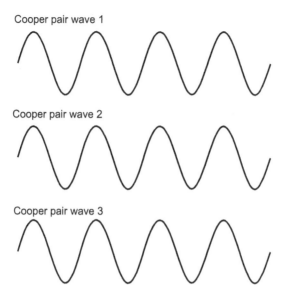

Figure 7.2   Coherence of waves. In a superconductor all the pair waves have the same phase and so these three waves should be superimposed: they are said to be phase coherent. A single wave function can then describe the entire collection of Cooper pairs.

it for many years. Fritz London, again displaying his deep insight, proposed, as part of a phenomenological theory of superconductivity, that the magnetic flux passing through the axial hole in a hollow, current-carrying, superconducting ring should be quantized in multiples of h/e (4.14×10^{-15} Weber); that is the flux can be zero, or h/e, or $2h/e$, or $3h/e$, and so on, but can have no value in between. This theoretical prediction that the flux must be a multiple of the basic quantum mechanical unit h/e is in complete contrast with classical physics, which would suggest that any current and magnetic flux can take any value at all. In his day, it was not possible to test this prediction because the available apparatus was not sufficiently sensitive to measure the small magnetic flux involved.

The advent of the BCS theory stimulated experiments to confirm the predictions made about flux quantization. It transpired that the results obtained provided convincing evidence for Cooper pairing. Careful experiments have verified that the magnetic flux is indeed quantized, but that the magnitude of the flux quantum is $h/2e$ (2.07×10^{-15} Weber): half London's predicted value. This is consistent with pairs of electrons, rather than single ones that London had originally assumed. Finding this gave an enormous boost to the acceptance of the BCS theory – being overwhelming evidence that the electrons are bound together in Cooper pairs: the result suggests that the charge on the current carriers is $2e$ (where e is the electronic charge).

Flux quantization is a direct result of the electron correlation and gives a further insight into the stability of persistent currents in superconducting rings. If a super-current is to decay, the flux must jump to another state with an integral quantum number: the system as a whole must be altered; many particles must change states

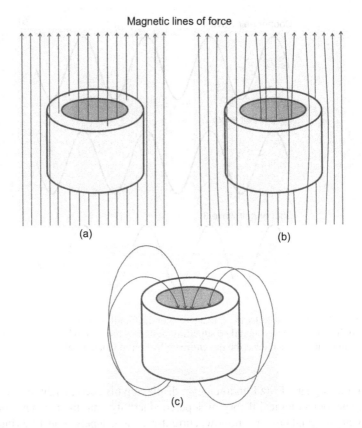

Figure 7.3 A ring in a magnetic field. (a) In the normal state: magnetic flux lines thread the ring and the metal itself. (b) When temperature is lowered below the critical temperature T_c, magnetic flux is ejected from the metal but still threads the ring. (c) As the outside magnetic field is removed, a current is induced in the ring which produces magnetic flux with the same value as before. The current and the magnetic flux persist and both are quantized.

simultaneously. The probability of this happening is vanishingly small because typical scattering processes in the solid affect only a few particles at a time. Hence supercurrents can and do persist. The nature of the flux quantization involved is a consequence of the zero resistance to a current flowing round a superconducting ring and the quantization of that current.

Let us take a ring made of a metal that can go superconducting, and place it in a magnetic field above the critical temperature, as shown in Figure 7.3(a). In the normal state magnetic field lines pass through the hole in the ring and also through the metal of the ring itself. If the ring is cooled below the critical temperature T_c, so that it goes into the superconducting state, the new situation shown in Figure 7.3(b) is produced: due to the Meissner effect, when the ring is made superconducting, the magnetic field is forced out of the superconducting metal of which the ring is made. However, flux still threads through the hole in the ring. If the external field is now removed, by Faraday's laws, as that field changes a current is induced in the ring that keeps the flux threading the ring at a constant value. This current persists round the ring. Hence the magnetic flux lines going through the hole remain trapped, as shown in Figure 7.3(c). Another novel feature is that both the current and the associated magnetic flux can only be increased in fixed steps. For the magnetic flux, drawn as a magnetic flux line, the fixed steps are now known to be $h/2e$. The **flux quantization** existing inside the ring can be visualized by fixing each magnetic field line with a value of $h/2e$ and ensuring that only an integral number of such lines can thread the hole in the ring in Figure 7.3(c).

The problem facing those experimentalists who planned to test this fundamental quantization of the current and flux in a superconducting ring is that the value of each flux quantum is extremely small: the amount of magnetic flux threading a tiny cylinder a tenth of a millimeter in diameter is about one percent of the earth's magnetic field. This makes experimental observation of flux quantization extremely difficult; to make a successful measurement, it is necessary to use very small rings. That was done. Flux quanta were first observed by measuring discontinuities in the magnitude of the magnetic field trapped in a superconducting capillary tube. In 1961 Bascom Deaver and William Fairbank at Stanford University, and at the same time Doll and Näbauer in Munich, Germany published papers in the same edition of *Physical Review Letters* reporting that they had been able to make sufficiently sensitive magnetic measurements to observe flux quanta and determine their value. The objective was to find out whether it is true that the magnetic flux threaded through a superconducting ring can take only discrete values. Both groups used very fine metal tubes of diameter only about $10\mu m$ (10^{-3}cm); then the creation of a flux quantum requires a very weak magnetic field of about 10^{-5}T, which can be handled experimentally provided that the earth's magnetic field is screened out. To carry out these superlative experiments, Doll and Näbauer used small lead or tin cylinders made by condensing metal vapor onto a quartz fiber. Deaver and Fairbank made a miniature cylinder of superconductor by electroplating a thin layer of tin onto a one-centimeter length of 1.3×10^{-3}cm diameter copper wire. The coated wire was put in a small controlled magnetic field, and the temperature reduced below 3.8K so that the tin became superconducting, while the copper remained normal. Then the external source of magnetic field was removed, generating a current by Faraday's law; as a result the flux inside the small superconducting tin cylinder remained unchanged, as shown in Figure 7.3(c). This tin cylinder now possessed a magnetic moment, which was proportional to the flux inside it. To measure this magnetic moment, a pair of tiny coils was sited at the ends of the tin cylinder and the wire wobbled up and down between them at about 100 cycles per second (rather like the behavior of the needle in a sewing machine). Then the magnetic moment was determined from voltage induced in the coils.

Deaver and Fairbank found that the flux was quantized, *but that the basic unit was only one-half as large as London had predicted on the basis of single electrons.* Doll and Näbauer obtained the same result (Figure 7.4 in which the states having 0, 1, 2 and 3 flux quanta can be seen). At first, this discrepancy from London's prediction was quite mysterious, but shortly afterwards the chief assumption of the Bardeen, Cooper, and Schrieffer theory that the superconducting electrons are paired provided the explanation of why it should be so. Everything had now come together.

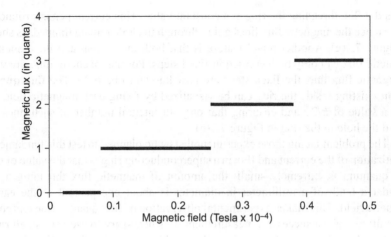

Figure 7.4 Experimental confirmation by Deaver and Fairbank of flux quantization in a tin cylinder having a very small diameter. The units of magnetic flux on the vertical scale are quanta of value $h/2e$ and the flux threading the cylinder can only take values on the lines shown, all the intermediate values not being allowed.

These experiments had verified the pairing postulate on which the BCS theory depended. All the Cooper pairs that carry the supercurrent are in the same quantum level. If Cooper pairs are caused to go into another quantum state, they must all change together. Any magnetic flux threading the hole in a superconducting ring can only exist as multiples of a quantum Φ_0, called the *fluxon,* given by

$$\Phi_0 = h/2e = 6.62 \times 10^{-34} \text{J s} / (2 \times 1.6 \times 10^{-19} \text{C}) = 2.07 \times 10^{-15} \text{ Wb}.$$

Here, the 2 in the denominator occurs because the electrons are paired. This value of a fluxon, equal to Planck's constant divided by twice the electronic charge, is extremely small. The tiny magnitude of this quantity can be put in perspective by noting that in the earth's magnetic field of about 2×10^{-5}T, the area that would be covered by a red blood corpuscle, which has a diameter of about 7μm, embraces roughly one flux quantum.

The existence of flux quantization also establishes the strict long-range phase correlation of the Cooper pairs with each other. A visual way of seeing this is shown in a simplistic fashion in Figure 7.5. A basic requirement of quantum mechanics is that a wave must be continuous and have only one value or it does not correspond to a single state. By analogy for a wave to travel round a loop of string, the string must not be cut and the ends must meet! Hence the coherent wave formed by superposition of the pair waves must complete an integral number of cycles round a ring (then the phase increases by an integral number of 2π once round the superconducting circuit). Addition of one more flux quantum into the bundle (initially containing an integral number n of flux quanta Φ_0 threading the ring) corresponds to an increase of exactly one more complete wavelength into the ring and the number of fluxons increasing to $(n + 1)$. Addition of two (rather than one) more flux quanta needs two extra complete

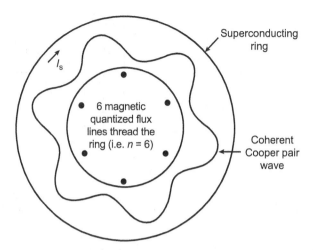

Figure 7.5 Relationship between flux quantization and phase coherence in a ring with a circulating supercurrent I_s. In this case 6 magnetic flux quanta thread the ring and there are 6 complete wavelengths round the circuit of the superconducting ring. If another flux quantum were to be added, there would then be 7 wavelengths round the ring. An intermediate state is not possible. Note the resemblance of this picture to that of the atom in Figure 5.2 but the huge difference in scale: the superconducting ring may be eight orders of magnitude (i.e. 10^8 times) larger – or even more – than the size of an atom. *Superconductivity is a giant quantum state.*

wavelengths, so that there are now $(n + 2)$ fluxons. So the observation of flux quantization also confirmed experimentally the existence of long range coherence as the large-scale quantum-mechanical behavior of electron pair waves in superconductors.

Flux quantization is not just restricted to a superconducting ring; that is, for a superconductor with a hole in it. Quantization always appears when a magnetic flux passes through any superconductor, such as those of type II, which are penetrated by an applied magnetic field in the form of fluxoids or bundles of fluxons (see Chapter 8).

As a first step in understanding the mechanism of superconductivity in the high T_c cuprates, it was vital to find out whether pairing of electrons is also involved. Several experiments have established that this is the case. One of the most compelling is the search for flux quantization carried out at the University of Birmingham by Colin Gough and co-workers shortly after the discovery of YBCO. They were able to measure the flux trapped inside a ceramic ring of YBCO. Their results are shown in Figure 7.6. They found that it is possible to excite the ring between its metastable quantum states corresponding to different integral numbers of the trapped flux. The multiphase nature of the YBCO allowed single quanta of flux to move easily in and out of the ring; this was crucial for the experiment to work successfully in the weakly superconducting material. They measured a value of $h/2e$ for the flux quantum Φ_0, hence establishing that *the superconducting electrons are also paired* in YBCO. In addition, this experiment shows that the long-range coherence of the pair waves is a property of high T_c cuprate superconductors, as it is for the more conventional materials. Hence flux quantization is a fundamental

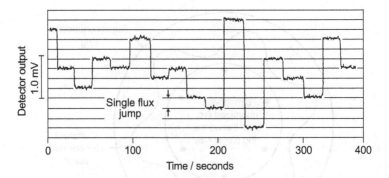

Figure 7.6 To measure the value of a flux quantum in a high T_c superconductor, a YBCO ring at 4.2K was periodically exposed (note scale on the graph abscissa is time) to a local source of electromagnetic noise causing the ring to jump between quantized flux states. The equally spaced lines shown here emphasize the quantum nature of the flux transitions because the flux jumps take place in integral numbers of a single flux quantum ($h/2e$). (Gough *et al.* (1987).)

property of all superconductors. Today, flux quanta play an important role in the many superconducting devices that depend for their operation on the so-called Josephson effects.

Josephson Effects

Age places no constraints on scientific work. While he was still a research student in the early 1960s at Trinity College, Cambridge, Brian Josephson put forward new fundamental ideas that completely changed the way in which superconductivity is viewed. The eminent American theoretical physicist Philip Anderson, who was a Visiting Professor at Cambridge at the time, has given a fascinating personal account of how Brian Josephson, then a young man of only 22 years of age, developed his far-reaching ideas and discovered the effects, which now carry his name.

Josephson considered what might happen when two superconductors are separated by a thin layer of an insulating material, which acts as a barrier to the flow of current. It had previously been recognized that, on account of their wave nature, electrons can tunnel through a thin insulating barrier between metals (see the discussion in Chapter 6 of how this phenomenon provided a powerful way of testing the BCS predictions). Tunnelling arises because the electron waves in a metal do not cut off sharply at the surface but fall to zero within a short distance outside. The electron wave leaks into the "forbidden" barrier region. Within this distance there is a small but finite probability that an electron will be found outside the metal. Therefore, when a piece of metal is placed very close (within about 10^{-7}cm) to another, electrons have a finite probability of penetrating the potential barrier formed by the insulating layer between the two metals. A small tunnelling current can be caused to flow across the junction by applying a voltage across the two superconductors (Figure 7.7).

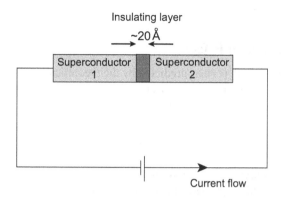

Figure 7.7 The principle of an experiment to test Josephson's prediction about tunnelling of Cooper pairs through an insulating junction between two superconductors.

In 1962, Josephson pointed out that, in addition to the ordinary single-electron tunnelling contribution, the tunnelling current between two superconductors should contain previously neglected contributions due to the tunnelling of Cooper pairs. The coherent quantum mechanical wave associated with the Cooper pairs leaks from the superconductor on each side into the insulating region. Josephson suggested that if the barrier is sufficiently thin, the waves on each side must overlap and their phases should lock together. Under these circumstances the Cooper pairs can tunnel through the barrier without breaking up. Thus, an ideal Josephson junction is formed between two superconductors separated by a thin insulating layer. The two electrons maintain their momentum pairing across the insulating gap and so the junction acts as a weak superconducting link. When there is a current flowing round a superconducting loop containing such a junction, there is a genuine supercurrent at zero voltage across the insulating layer. Josephson provided an equation for the tunnelling current (see Box 8). He was much puzzled at first as to the meaning of the fact that this current depends on the phase, which may in part explain the title of his paper reporting his work: "Possible New Effects in Superconductive Tunnelling" in the newly created journal *Physics Letters* rather than the prestigious *Physical Review Letters.*

Anderson returned home to Bell Telephone Laboratories extremely enthusiastic about what Josephson had done and eager to confirm pair tunnelling experimentally. He told a colleague John Rowell of his conviction that Josephson was right. Rowell pointed out that he had noticed suggestive things in tunnelling experiments that he had made on superconductors which could have indicated that he might actually be seeing Josephson effects. Motivated by the new ideas, he set off to study a new batch of tunnel junctions. In those days it was not easy to see the effects but he proved able to do so. Anderson and Rowell had a number of advantages going for them. In the first instance, Rowell's superb experimental skill in making good, clean, reliable tunnel junctions was especially valuable. The direct personal contact with Josephson ensured that they knew what to look for and could understand what they saw. When they came to publish their findings, they were rather more confident than young Josephson had been: now that they had understood and extended the theoretical ideas

Box 8

The dc Josephson Effect

The tunnelling current (I) was predicted by the theoretical work of Josephson to be given by the famous dc-Josephson relation, which is central to a detailed understanding of superconductivity:

$$I = I_C \sin(\theta_1 - \theta_2). \tag{1}$$

Here I_C is the maximum supercurrent that can be induced to flow, that is the critical current. This expression, also sometimes known as the sinusoidal current phase relationship, relates the current I to the phases θ_1 and θ_2 of the pair waves on each side (1 and 2) of the junction. The external current drives the difference $\Delta\theta$, equal to $(\theta_1 - \theta_2)$, between the phases of the macroscopic waves travelling in the superconductors on opposite sides, as illustrated in Figure 7.8 and defined by equation (1).

A second Josephson equation applies when an ac voltage is applied across the junction. Then the phase difference $\Delta\theta$ increases with time t as:

$$d(\Delta\theta)/dt = 2eV/h. \tag{2}$$

This effect allows a Josephson junction to be used as a high frequency oscillator or detector.

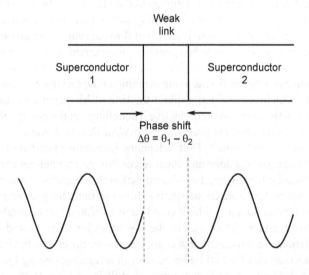

Figure 7.8 Tunnelling of a Cooper pair wave across the weak link between two superconductors. The phase shift $\Delta\theta$, equal to $(\theta_1 - \theta_2)$, across the barrier is shown diagrammatically by the two waveforms. It is not physically possible to indicate a waveform in the barrier itself.

they were able to change the title of their paper from "Possible ..." to "Probable Observation of the Josephson Superconducting Tunnelling Effect", which they published in *Physical Review Letters.*

Anderson has recalled that it was no coincidence that Josephson carried through these developments in the stimulating atmosphere of the Cavendish Laboratory. Not only did Josephson make the all-important theoretical leaps but he also explained how to observe the effects that he had predicted. Entirely by himself, he solved the Cooper pair tunnelling problem and its physical ramifications in a complete and rigorous manner.

A patent lawyer, consulted by John Rowell and Philip Anderson at Bell Telephone Laboratories, gave his opinion that Josephson's paper was so complete that no one else was ever going to be very successful in patenting any substantial aspect of the proposed effects. That was not to say that patents pertaining to working applications could not be made.

At first sight, the Josephson effects would appear to be just an esoteric part of fundamental physics far removed from the "real world" of work and business. Yet they now have applications in numerous areas (see Chapter 9), which must have been inconceivable at the start. This is one of the many examples of the spin-off from research into the fundamental properties of nature such as superconductivity.

8 Applications of Superconductors: Enabling Technologies of the 21st Century

Almost from the moment that Kamerlingh Onnes first realized that a superconductor has no electrical resistance, he appreciated the potential for producing commercial magnets. He hoped that it might be possible to generate powerful magnetic fields by passing high currents through superconducting wire coils (usually called solenoids). The zero resistance of a superconducting solenoid avoids the severe drawback of large Joule heating in coils made from resistive metals. Once the magnetic field has been set up in a coil of superconducting wire carrying a persistent resistanceless current, it needs no further electrical power to maintain it: there is no energy loss.

To sustain a field of a few Teslas in an electromagnet with conventional resistive wire coils would require using a sizeable fraction of the power consumed by a small city. What is more, to prevent melting the coils, all of the heat produced by Joule heating due to the resistance of the wire must be removed by a continuous flow of as much as a thousand liters per minute of cooling water. Clearly a conventional electromagnet is large and costly to run. By contrast, negligible power is needed to operate a superconducting magnet producing a similar field, although it must be kept cold to ensure that the coil does not go normal.

In principle, superconductors offer a much cheaper solution to producing high field magnets. Unfortunately Kamerlingh Onnes soon discovered that the fields, which could be generated before his superconducting metals were driven normal, were much too small to be of practical use. This was because the critical magnetic fields that destroyed superconductivity were very small for the superconductors that he had found. His work was on elements that were later recognized to be typical type I superconductors. He did discover that lead is a superconductor with a comparatively high critical temperature (7.2K) and a relatively large critical magnetic field but even that is only 0.08T. Nevertheless, lead has found many uses in superconducting devices over the years.

Superconducting technology can be broadly divided into two categories: large-scale applications, usually requiring the use of long lengths of wire or tapes, and small-scale electronics often utilising thin films. This chapter is devoted to the large-scale applications while the small-scale ones are discussed in Chapter 9. Prominent

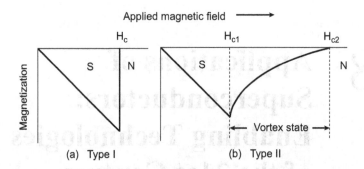

Figure 8.1 (a) Magnetization versus applied magnetic field for a type I superconductor, which exhibits a complete Meissner effect (that is shows perfect diamagnetism). As the applied magnetic field is increased, it is completely rejected from the interior of the super-conductor. Note that to correspond to this diamagnetic behavior, the magnetization is plotted in the negative direction on the vertical scale. Above the critical field H_c the specimen is a normal conductor and the magnetization is too small to be seen on the scale shown. N refers to the normal state and S to the superconducting state. (b) Magnetization curve for a type II superconductor. The flux starts to penetrate the specimen at a field H_{c1} and penetration is completed at H_{c2}. The material is in a vortex state between H_{c1} and H_{c2} and it can carry super-current up to H_{c2}. Above H_{c2} the material is a normal (N) conductor.

among the potential large-scale applications is the use of superconductivity in the electrical power industry both for generation and distribution. Discovery of a new type of superconductor since the days of Kamerlingh Onnes has made the production of high magnetic fields possible.

Magnetic Properties of Type II Superconductors

A considerable advance was made throughout the 1950s and 1960s with the discovery (described in Chapter 2) of type II superconductors, including Nb_3Sn and NbTi, which can sustain much higher magnetic fields than type I without going normal and can be made (but not without some difficulty) into wires. By using these materials it became possible to construct a wide range of compact magnets capable of generating fields up to several Teslas and about ten times cheaper to run than conventional magnets. Such magnets have formed a substantial part of a small but flourishing industry based on superconductivity.

All superconductors show the Meissner effect at sufficiently low fields and completely expel magnetic flux from their interior. Since a type I superconductor is perfectly diamagnetic, when an increasing magnetic field is applied, the flux density within the specimen remains at zero (if we neglect the thin penetration depth at the surface) because flux exclusion is complete up to the critical field H_c. The way in which the sample magnetization varies as a magnetic field of increasing strength is applied is shown in Figure 8.1(a). When the applied field strength reaches the critical value H_c, the superconductor is driven into the normal state, and the magnetic flux inside is no longer zero. At higher applied field strengths the material behaves as a normal metal.

Many alloys show type II superconducting behavior that is marked by partial flux penetration in moderate magnetic fields. As far back as the late 1930s, Shubnikov had been able to establish that the superconducting properties of such alloys in a magnetic field were fundamentally different from those of metallic elements (see Chapter 2). He found that there are three distinct regions of behavior in type II superconductors with increasing applied magnetic field; these are illustrated in Figure 8.1(b). Up to a fairly low magnetic field H_{c1}, the magnetic flux is totally excluded in a similar manner to that observed for a type I superconductor. However, above this lower critical magnetic field H_{c1}, partial flux penetration occurs, although the bulk of the material still remains superconducting. There is a limit: when a magnetic field larger than a much higher, upper critical field H_{c2} is applied, all of a type II material becomes a normal conductor. In the range between the lower critical field H_{c1} and the upper one H_{c2} a type II superconductor is a mixture of superconducting and normal regions – variously known as the intermediate, mixed or vortex state. The important feature of a type II superconductor, so far as technological application is concerned, is that in this intermediate region a zero electrical resistance path still exists while there can be a large magnetic flux present in other parts. For some type II superconductors, the magnetic field H_{c2} required to destroy superconductivity completely can be as high as several Teslas. By taking advantage of this high field that these type II superconductors can sustain, it is possible to produce compact high field superconducting magnets.

The structural features of a type II superconductor in the mixed state containing both superconducting and normal regions are particularly revealing. When a magnetic field, larger than the lower critical value of H_{c1} but less than H_{c2}, is applied to a type II superconductor, it penetrates the sample as discrete flux lines, which form tubular regions parallel to the applied field, as shown in Figure 8.2(a). These tubes, threaded by the magnetic flux, are normal regions (Figure 8.2(b)). They are surrounded by a matrix of superconductor, which can still carry supercurrent. Each tube contains the smallest possible amount of flux, namely a flux quantum (see Chapter 7). The magnetic flux in each tube is sustained by a vortex of persistent current circulating in the superconducting region around the normal core. The existence of these magnetic flux tubes was first predicted by the Russian theoretician Alexei Abrikosov in 1957; he showed that energetically it is favorable for the flux tubes to form a regular array, called the vortex lattice, shown in plan in Figure 8.3.

There was some initial skepticism as to whether these ideas about flux tubes put forward by Abrikosov were indeed valid. Doubts were scotched by some elegant work by Uwe Essmann and Hermann Träuble at the University of Munich in which they showed that it was possible to "see" and photograph quantized vortices. Their technique was based on earlier observations of domains in a ferromagnetic material revealed by "decorating" the array of magnetic field lines with small magnetic particles; the resulting patterns formed by the particles could then be photographed. The vortices in a superconductor are very much smaller than domains in ferromagnets so that decorating them was more difficult. Iron atoms from a vaporized source were allowed to condense onto the end surface of a superconductor. It was

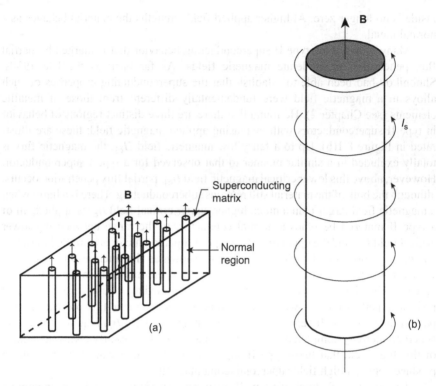

Figure 8.2   (a) The vortex lattice formed in a type II superconductor comprised of tubular regions parallel to an applied magnetic field larger than the lower critical value of H_{c1} (but less than H_{c2}). (b) Each tube is a normal region (shown shaded) surrounded by a matrix of superconductor, which can still carry supercurrent. Each tube is threaded by a flux quantum sustained by a vortex of persistent current $\mathbf{I_s}$ circulating in the superconducting region around the normal core.

observed that iron congregated at the vortices. Examination under an electron microscope at room temperature then showed the triangular array of vortices (Figure 8.3) predicted by Abrikosov.

   The development of the normal cores stabilizes the mixed state of type II superconductors by lowering the energy. The conditions which determine whether a superconductor is type I or II are outlined in Box 9.

   When a current is passed through a type II superconductor in the mixed state having these normal tubes containing magnetic flux, a force of the type first recognized by Lorentz can cause the flux lines to move. Any such movement would result in dissipation of energy and runs counter to the central aim of using the superconductor to carry current without energy loss. An important advance was made when it was found possible to prevent the vortices from moving by fixing them on "pinning centers" such as dislocations, grain boundaries or at defects formed by cold working the material. These materials, in which pinning centers inhibit flux line movements, are called *hard superconductors;* these are the ones actually used in high magnetic field applications such as commercial magnets.

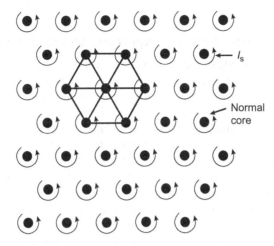

Figure 8.3 Plan view of the vortex lattice in a type II superconductor in a magnetic field. The applied magnetic field comes out of the plane of the diagram. The normal core regions through which the magnetic flux **B** penetrates are cylindrical, as shown in Figure 8.2, and are arranged regularly within the surrounding superconducting matrix as indicated by the hexagonal grid. Each is sustained by a vortex of persistent current I_s at its interface with the superconducting matrix.

Type II Superconducting Wires and Magnets

To use a superconducting coil as a magnet, usually requires that driving current can be made as large as possible, the upper limit being set by the critical current which destroys superconductivity. For safe use it is essential that the coil does not breakdown. Should a high field superconducting magnet suddenly go normal, the resulting heat generated would cause the liquid helium coolant to boil off rapidly – with frightening consequences! Catastrophes were by no means unknown in the early days. Secure pinning of vortices is not sufficient for safe and reliable operation of wires made from hard superconductors. Local sudden disturbances in current, temperature or magnetic field can cause instabilities, known as flux jumps, to occur. Entire bundles of flux can suddenly break away from their pinning centers and migrate through the material – inducing resistance and creating a hot spot. If a region of normal resistance is developed, further local heating occurs there, causing an avalanche effect until the complete coil becomes unstable and goes normal – very rapidly. To avoid this, the energy stored in the magnetic field must be dissipated quickly. These problems, which presented a considerable obstacle to the early development of superconducting high field magnets, were overcome around 1965 by using "*stabilized wires*". To make these, the superconductor is coated with a normal conductor such as copper or aluminium that has low resistance to flow of both electrical current and heat. This coating provides a good normal conducting path: if an instability due to a flux jump occurs, heat can be rapidly conducted away, the hot spot cools down and becomes superconducting again.

Box 9

What determines whether a superconductor is type I or II?

There are two fundamental parameters, which determine whether a superconductor is of type I or type II. First is the penetration depth λ to which a magnetic flux penetrates into the surface of a superconductor; this is the skin in which current flows (Chapter 2). Second is the coherence length ξ – corresponding to the smallest distance over which the electronic properties of a superconductor can change (Chapter 6). These two parameters operate together at any boundary between a normal and superconducting region. The reason why the mixed state of a type II superconductor can exist stems from the presence of interfaces between normal and superconducting regions; the overall energy is lowered at these interfaces due to a competition between an energy increase due to magnetic field penetration and a decrease caused by the formation of Cooper pairs within the coherence length. Whether a material is type I or II is determined by the Ginsburg–Landau criterion: whether the ratio λ/ξ of the penetration depth λ to the coherence length ξ is smaller or greater than $1/\sqrt{2}$. For type I superconductors λ/ξ is less, whereas for type II superconductors λ/ξ is greater. The physical difference between the two types stems from the fact that it is energetically favorable for a type II superconductor in the mixed state to form as many boundaries between superconducting and normal regions as possible; this is not the case for type I. Materials, such as metal alloys, tend to have a longer penetration depth than the coherence length, that is λ/ξ is greater than $1/\sqrt{2}$. This results in a negative energy near the surface which is conducive to the formation of normal-superconducting boundaries – so these materials are type II.

We have known since the late nineteen twenties that quantum physics can predict the behavior of matter on very small scales (see Chapter 5). What is so significant about superconductors and superfluids is that they show us amazing effects at a far larger scale where they can actually be seen (Chapter 7). So it is easy to appreciate why interest in the fundamental large-scale quantum effects found in superconductors and superfluids continues to be at the cutting edge of science. Belated recognition has at last been given to two Russians, Alexei Abrikosov (now at the Argonne National Laboratory in Illinois) and Vitaly Ginzburg, retired head of the theory group at the P.N. Lebedev Physical Institute in Moscow, for their theoretical explanation of type II superconductors. They share the 2003 Nobel Prize in Physics with Tony Leggett for his pioneering work while at the University of Sussex on macroscopic quantum effects in superfluid helium.

Niobium–titanium alloy is commonly used to make high field magnets. One way of making a stabilized wire is to incorporate a niobium–titanium rod into a copper billet and then draw the composite down to wire dimensions. Multicore wires containing many fine superconducting threads embedded by co-drawing in a copper or copper–nickel matrix are still used commercially to construct magnet coils but the future lies in using high T_c materials (Figure 8.4).

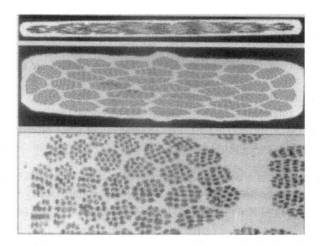

Figure 8.4 Cross sections of high temperature superconductor multicore wires. At the top is that of a Bi2223 tape of dimensions 4.1mm × 0.2mm. In the middle is YBCO tape of dimensions 3mm × 0.6mm comprised of 36000 filaments. At the bottom is a magnified view of a bundle of filaments. The tapes were made by the American Superconductor Corporation. (After J. Tallon (2000).)

The Challenge of Commercial Fabrication of High T_c Superconducting Ceramics

The discovery of the ceramic cuprate superconductors, especially those with critical temperature T_c greater than the normal boiling point of liquid nitrogen (77K), immediately focused attention on the possibility of widespread technological applications. In principle, the completely frictionless flow of electrons is an electrical engineer's dream. If it was possible to produce an inexpensive superconductor operating at room temperature which was easy to manufacture and could be readily fabricated into wires and films, there is no doubt that there would be a technological revolution. However, outside laboratories, hospitals or industrial concerns, a superconducting environment is largely a pipe dream and may well remain so for many years to come.

Nevertheless, the last few years have seen enormous advances and the prospects for widespread applications in the next decade or two look hopeful. It should be emphasized that the arrival of the high temperature superconductors does not alter the basic engineering physics on which applications are based. But a higher operating temperature does have distinct advantages. Cooling by liquid nitrogen (boiling point 77K) is about 50 times cheaper than with liquid helium and the associated cryogenic hardware is much less complex. So working with liquid nitrogen is much easier and cheaper than operating in liquid helium just above the absolute zero. The need for costly cryostats and insulating materials is largely eliminated. An additional advantage is that the latent heat of vaporization of liquid nitrogen is some seventy times greater than that of liquid helium. Hence, any device cooled to liquid nitrogen temperature will remain cold for a very much longer time than that cooled by the same volume of liquid helium. Furthermore, it is possible to make semiconducting devices that operate at 77K; this enables the development of powerful new hybrid semiconductor–superconductor systems. A wide range of applications in the electronics field is in the offing (see Chapter 9).

The cost of suitable bulk ceramic superconductors for applications is a concern of all potential users. At present bulk YBCO costs rather less than 3 dollars (£2) per gram, but the price is expected to fall to as low as 0.5$, as the market demand increases.

Apart from material cost, designs are needed that reduce the quantity used. Necessity drives: there are many clever design innovations. One typical industrial problem is to make large pieces of ceramic of complex shape. To achieve this, an approach akin to welding shows promise. Instead of making one large specially shaped piece, several small sections can be joined. This has required invention of a method for "glueing" YBCO to YBCO. This has been achieved by Boyd Veal and collaborators at Argonne National Laboratory using an intervening layer of the thulium-based copper oxide superconductor (TmBCO), which has a lower decomposition temperature than that of YBCO, although its crystal structure is the same and its lattice parameters are closely similar. The parts to be joined and the TmBCO spacer are heated to a temperature intermediate between the two decomposition temperatures. During cooling, the YBCO pieces act as seeds for the thulium based spacer material. Mechanically strong joints can be produced in this way with a superconducting current-carrying ability entirely comparable with that of the undisturbed bulk material. Commercial fabrication costs can be much reduced by this "welding" approach.

In the euphoria following the discovery of the high temperature super-conductors, it was widely believed that they would rapidly have a considerable impact in superconducting technology and supersede the earlier type II materials used in coil windings. The reality has been somewhat different; progress in utilising high temperature superconductors in high magnetic field applications has been slower and much more difficult to achieve than initially envisaged. While the values of both critical temperature T_c and magnetic flux B_c for new materials are favorable, it was found early on that the values of critical current density J_c that destroyed superconductivity in the coils were by no means as high as could be hoped for. For most applications a critical current density J_c value of 10^5 to 10^6 amps/cm^2 is necessary for superconducting magnets to become commercially viable and outperform conventional electromagnets. Initially, values only up to about 10^3 amps/cm^2 could be reached at a temperature of 77K and to make matters worse these decreased rapidly in a magnetic field. After much effort, current densities have now been improved substantially: up to or better than 10^5 amps/cm^2 in fields of several Teslas.

The reasons for the low achievable values of current density lie in the microscopic structure and nature of ceramic high temperature superconductors. To optimize the properties of YBCO, attention must be paid to the conditions of its manufacture. For example, careful oxygenation is necessary. A relatively small change in the oxygen concentrations can have a dramatic effect on the critical temperature T_c (Figure 4.6) and also critical current J_c (or even change the material from being a superconductor to an insulator). Degraded regions inside super-conductors can exist. A further problem is caused by the anisotropic nature of the cuprates. As a consequence of the layer-like structure (Chapter 4), current flows readily along the $a-b$ plane in a single crystal of YBCO but much less easily, by at least one and two orders of magnitude, normal to the plane. In a polycrystalline material in which the grains are arranged randomly, the current tends to avoid those grains that have their c-axis lying in the direction of the current flow. Hence, the net current density is greatly reduced compared with that expected for a single crystal. Weak coupling across grain boundaries of polycrystalline samples of ceramic super-conductors is another major factor resulting in the low observed values of critical current J_c. Extensive research carried out at the IBM Laboratories in Yorktown Heights has shown that a reduction in critical current J_c of about two orders of magnitude can result, if the grain boundaries are misorientated by more than a few degrees. Overcoming this weak coupling, or granularity, problem has required some sophisticated materials science. Larger values of critical current can be obtained by texturing the samples so that the orientation between adjacent grains is greatly reduced.

In the working conditions of a superconducting magnet, both a high current and a high magnetic field exist simultaneously. The high T_c cuprates show behavior akin to that of type II superconductors (Box 10). In the working conditions which occur in high field magnets, these materials exist in a mixed state with a vortex lattice. As discussed earlier in this chapter for type II superconductors, in the mixed state magnetic flux lines in the superconducting coil experience a Lorentz force, which can cause them to move; such flux line movement would result in the dissipation of energy and corresponds to a resistance to the flow of electrical current. Therefore one requirement for obtaining high values of critical current J_c is that flux lines do not move. To prevent this movement, it is essential to pin the flux lines securely. The earlier technology based on type II hard superconductors was viable because high critical current values could be produced in those materials by ensuring that they contained large numbers of crystal defects suitable for pinning the flux lines securely. More recently, industrial companies in particular have spent a vast amount of effort and resources in attempts at making high T_c superconducting wires, in which flux pinning is sufficiently strong for high critical current values to be possible. Flux pinning is a parameter that is especially sensitive to the metallurgical state of a particular material and its defect population. In addition to flux flow, losses arise due to flux creep whereby a vortex or bundle of vortices can jump or "hop" between two adjacent pinning centers. The purpose of using the cuprate superconductors is to operate at the reasonably high temperature of 77K; however, as the temperature is increased, thermal effects become increasingly important. Flux flow and creep can be rapid at high operating temperatures. Overcoming such problems remains particularly important for the high temperature superconductors.

The production of high magnetic fields, and also applications in the power industry both for generation and distribution of electricity, requires the fabrication of high quality superconducting wires, tapes or cables, which presents a formidable

Box 10

The high T_c superconductors also show type II behavior

The high T_c superconductors show type II behavior. In these layer-like materials the coherence length, ξ, is highly anisotropic and exceptionally small. For example, in YBCO in the low temperature limit, the coherence length is 1.4nm in the $a-b$ plane and only 0.2nm in the c-direction. This can be compared with a coherence length of about 40nm for niobium. The penetration depth, λ, of high T_c cuprates is relatively large, that for YBCO is 140nm in the $a-b$ plane and 700nm in the c-direction compared with a value of around 40nm for niobium. Thus YBCO, and the other cuprate superconductors, can be regarded as extreme forms of type II superconductors, obeying the condition that λ/ξ is much greater than 1. They are also in the so-called "clean limit" where the coherence length is much shorter than the electron mean free path at T_c (which is about 10 to 20nm).

The cuprates show a vortex lattice. The superconducting CuO_2 layers are weakly interacting and this gives rise to loosely connected vortex lines, which are easily made to vibrate by thermal fluctuations. This can give rise to a completely new physical phenomenon: sufficiently strong vibrations can cause the vortex lattice to "melt" and change into a vortex "liquid". The behavior is analogous to the effect of thermal vibrations in causing ice to melt into water. Random fluctuations in the vortex liquid can give rise to electrical resistance. In itself, vortex lattice melting into a "liquid" is currently a hot topic.

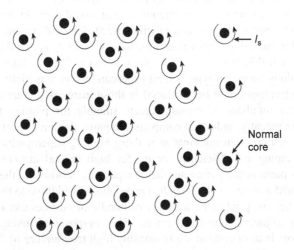

Figure 8.5 A disordered vortex glass formed in a highly defective material. The arrows represent the vortices of persistent current circulating in the superconducting region around the normal cores. The disorder can be contrasted with the regular vortex lattice formed in a nearly defect free material and shown in Figure 8.3.

What happens when a vortex liquid freezes? If the superconductor has few impurities or defects, the vortex liquid freezes into a regular triangular lattice in the usual way found for type II materials and first suggested by Abrikosov. However, if, as in cuprates, the material contains a large number of impurities or defects, the flux lines can "freeze" in a disordered manner giving rise to what is known as a "vortex glass" as shown in Figure 8.5. The idea of a vortex glass was first put forward by Mathew Fischer and colleagues at the IBM company and it has analogies with the spin glass mentioned in Chapter 11. At first, the suggestion that a vortex glass might exist was greeted with considerable scepticism, but careful measurements have shown that it occurs in cuprates. Here is yet another area of new fundamental physics emerging from the study of these exotic materials.

---

technological challenge. An enormous amount of work has been carried out in producing yttrium and bismuth based cuprate superconducting wires; there is also some literature on thallium and mercury based wires. The last decade or more has seen considerable progress in overcoming the fabrication problems and several companies have been able to produce good quality prototype superconducting wires in kilometer lengths. Because of the very brittle and non-ductile nature of the ceramic oxides, they must be sheathed or encased within a suitable metal that conducts heat and electricity well. Silver has been found to be preferable to other metals. Unfortunately copper or aluminium, the obvious choice of materials in terms of cost and availability, have very unsuitable phase equilibrium properties when in contact with copper oxide perovskite at the temperature at which the superconductor must be processed. This is not the case for silver. In addition, oxygen can diffuse easily through this metal, greatly benefitting the manufacturing process, as well as assisting in the stabilization and grain alignment of the oxide superconductors.

To make wires, a thoroughly mixed powder of the constituent components in proportions needed to make the cuprate superconductor is tightly packed into cylindrical silver billets. These are then repeatedly drawn and swaged until they form filaments about 1mm thick. Bundles of these filaments are then rolled together to form a tape which is some 4–6mm wide and less than 1mm thick (Figure 8.4). The final step is to wind a fixed length of tape into a spool and anneal for several days in an oxygen atmosphere. Naturally, details of the production processes are obscured by commercial secrecy. At the moment one of the most promising candidates for large-scale applications are the bismuth based ceramics contained within a silver sheath. Useful values for the current density can be obtained with this composite – largely because it is possible to achieve a high degree of grain alignment, an essential requirement. This is a decisive advantage of BSSCO over YBCO for which grain alignment is much more difficult. Other factors, which can reduce the critical current, are discussed later in this chapter. To date, wires of length running to several kilometers have been produced from bismuth based ceramic superconductors (BSCCO) encased in silver by companies such as the American Superconductor Corporation in the United States and the Sumitomo Electric Corporation in Japan.

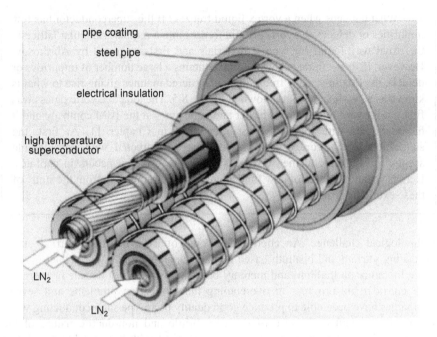

Figure 8.6 Design for a cable comprised of spirally wound high T_c superconducting tapes. Liquid nitrogen flows in ducts along the cores. (After J. Tallon (2000).)

Prospects for a Superconducting Grid Network?

There are exciting prospects for future widespread use of superconductivity in the electrical generation and distribution industry. Each of the components of an electric power system can be replaced by a superconducting equivalent. The most immediately obvious is to convey electricity by zero resistance cables. After generation, electricity is transmitted in the grid either by overhead lines or by underground cables. Overhead high voltage transmission lines on pylons are considerably cheaper than underground cables but suffer from several severe drawbacks. They are an eyesore and unacceptable in areas of outstanding natural beauty. In addition, they use up valuable land, can be damaged by lightning strikes and vandals; there is also an ongoing controversy as to whether people in the close vicinity of such power lines can suffer from harmful exposure to the low level electromagnetic radiation. Putting the cables underground eliminates many of these problems but the cost can be prohibitive. Conventional underground cables suffer energy losses amounting to several percent due to the Joule heating from the resistance of the wire. Modern underground power cables are technologically very complicated requiring the use of elaborate and expensive cooling. Making power transmission cables from high temperature superconductors would overcome some problems. Research and development of superconducting cables continue at a great pace. Prototypes are already in existence and it is likely that those for commercial use will appear in the early years of this millennium.

In cables the superconductors in the form of a flat tape are non-inductively wound in a helical manner on top of a copper former. To relieve stress, this is covered with a layer of black carbon tape and then insulated using a suitable dielectric such as a laminated polymer material. A typical design for a high temperature superconducting cable operating at 77K, the normal boiling point of liquid nitrogen, is shown in Figure 8.6. Any design requires at least two cables in order to transport the current to and from the load. Each conductor is cooled by liquid nitrogen flowing through the center of it. The superconductors are immersed in a liquid nitrogen return line which may be vacuum insulated from the outside world and encased in a shield of stainless steel and an outer thermal shield. Cryogenic cooling is produced by some kind of Stirling cycle engine, housed in a series of sub-stations. Liquid nitrogen is circulated along the cable by turbo-pumps to a second sub-station, which can be situated several kilometers away. At the second sub-station, the nitrogen is collected, cooled, and pumped back to the first station. Each sub-station contains the vacuum pumps required for maintaining the thermal insulation.

One advantage of superconducting cables over conventional underground oil cooled cables is considerable space saving. Many more superconducting cables with much higher current density ratings can be laid in the existing ducts used for conventional cables. This should have enormous benefits in densely populated inner city areas such as New York and Tokyo, enabling a far greater quantity of electricity to be distributed by the existing cable network. Finally, looking to the future, superconducting transmission may well become viable in the next generation of space stations.

Superconductivity already plays a role in other aspects of electrical generation and distribution such as transformers, fault current limiters, generators and motors.

Transformers using high temperature superconducting wires may well replace those using normal windings since they can be made both more efficient and more compact. Most of the energy losses in conventional transformers are due to resistive heating of the windings. This can be eliminated by using superconducting wires, although there are still small alternating current losses. Such wires can carry between one and two orders of magnitude higher current density than conventional windings – resulting in greatly reduced weight and size. A disadvantage of conventional transformers is their use of coolant oil which is both a fire hazard and a potential contaminant should it leak. By contrast, the superconducting transformers use liquid nitrogen which is relatively harmless. In conventional power transformers a major problem is heat degradation of the insulation. This can seriously reduce the lifetime of the transformer if the power rating is slightly exceeded causing a temperature rise of only a few degrees for a short period of time. Again, these problems are eliminated by operating transformers at the boiling point of liquid nitrogen.

Any short circuit within a power line can produce a rapid reduction in the impedance of the power grid and a corresponding current surge, known as a fault current. A short circuit can arise accidentally from a lightning strike or a tree branch hitting the lines. An important requirement for power transmission lines is to be able

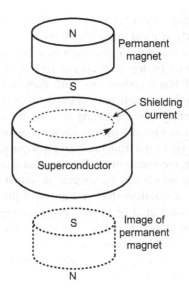

Figure 8.7 Levitation. If a magnet with its south pole down is moved towards a superconductor, the Meissner effect induces a shielding current with an equal but opposite associated magnetic field with the south pole up. The magnet is repulsed by this mirror image magnetic field.

to provide effective protection against such events. Techniques have been developed for dealing with fault currents but so far none have proved to be totally satisfactory. An ideal fault current limiter should have zero impedance during normal operation but show a very large impedance under overload conditions. It should also provide rapid detection and response to a current surge and then be able to recover rapidly. In addition, it needs to be extremely reliable, automatic, compact, lightweight and inexpensive. Superconductors have always been considered to be particularly suitable for fault current limiters since they are able to change rapidly from the zero impedance superconducting state to the high impedance normal state. When used in conjunction with suitable power electronics, a superconducting current limiter can rapidly detect a sudden surge; it can take appropriate action and then return almost immediately, within less than a cycle, to the normal mode of operation after the fault is cleared. The discovery of the high temperature superconductors has given considerable impetus to development of fault current limiters because they are more compact and reliable, to say nothing of being cheaper, than the earlier prototypes which used type II superconductors.

There has also been considerable effort to promote use of high temperature superconductors for power generators and motors. Superconducting motors are smaller and more efficient than conventional ones. Again a decisive factor is the higher current density that can be carried by superconducting windings compared with conventional copper ones. In the U.S. it is estimated that nearly 60% of all the electricity produced is used to run electric motors for pumping oil, gas, water and air as well as running compressors and a whole variety of machinery for industry, offices and the home. Replacement of these motors by smaller more reliable and cheaper superconducting ones would save considerable electrical power.

Another developing area in the power industry is the superconducting ac generator which can convert mechanical energy to electrical energy more efficiently than any other way. In addition, such machines can generate power at transmission voltages, eliminating the need for special transformers at generating stations. Several generator programs using high temperature superconductors are in place around the world.

Frictionless Bearings and Flywheels

An intriguing application of superconductors is in constructing virtually frictionless bearings based on levitation in a magnetic field. Frictionless bearings, which operate at liquid nitrogen temperatures, can now be made from high T_c cuprates, although extending the range to room temperature remains a mechanical engineer's dream.

Magnetic bearings have lower power losses than mechanical ones; those that use ceramic superconductors offer much promise because they have very small losses. The simplest form of a superconducting bearing consists of a small but powerful permanent magnet levitated above a superconductor (Figure 8.7). This is the same set up as that described earlier for demonstrating the Meissner effect (Figure 2.5). The exclusion of magnetic flux by the superconductor induces a shielding current with an associated magnetic field that is exactly equal but opposite to that of the permanent magnet. Accordingly the magnet faces its mirror image and is held above the superconductor by repulsion from that magnetic image. The closer the magnet is placed to the superconductor, the stronger the repulsion; the farther away, the weaker the force. Balance levitates the magnet. In general, a system of permanent magnets interacting with each other is unstable. (This is Earnshaw's theorem recognized as early as 1842.) Superconductors present a rare exception to this theorem due to their ability to carry current without loss. In practice, it is the presence of flux pinning centers that allows stable levitation to take place. Magnetic flux lines passing through the pinning centers tend to become trapped there (Figure 8.8). The flux lines between the superconductor and the magnet act like "mechanical springs" attached at each end. If the floating magnet attempts to move, it is pulled back into position – producing stable levitation.

The simplest design for a bearing is to attach a rotor to the magnet. Ceramic YBCO is a suitable material for superconducting bearings operating at the temperature of liquid nitrogen. At the moment it is possible to obtain levitation pressures of some 300kPa, which are enough to provide some spectacular demonstrations such as one from Japan showing a floating huge sumo wrestler.

A natural extension of the successful development of near frictionless bearings using high temperature superconductors is their employment in flywheels. There is renewed interest in the use of flywheels for energy storage over short periods of time. The potter's wheel is one of the earliest examples of the use of a spinning wheel for storing energy. A flywheel rotor possesses, and hence can store, kinetic energy by

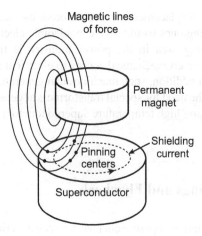

Figure 8.8 Flux pinning on defects in a superconductor. Small vortices of current flow round pinning centers trapping the quantized flux lines.

virtue of its rotational speed. Storing kinetic energy in a flywheel is particularly useful because it can readily be converted into electrical energy and vice versa. Bearings are needed to support the flywheel rotor. One limiting factor for the design of flywheels is the energy losses in the bearings; these can be almost entirely eliminated by the use of superconductors. YBCO is used because it can remain superconducting in a much higher magnetic field at 77K than BSCCO can. Since the superconducting part of the bearing must be maintained at 77K (or below), the simplest design is to keep the superconductor and its refrigeration system stationary. In this case, the ambient temperature permanent magnet is rigidly connected to and rotates with the flywheel rotor.

Nowadays, the power sector has begun to use flywheels. Inexpensive energy storage enhances the economics of both wind-powered and nuclear generation of electricity. Storage enables electricity to be generated during one period (for example when the wind is blowing) and consumed at another time, when a customer wants it. Storage may also improve power quality by smoothing out fluctuations, such as voltage dips and short term supply surges as customers switch in and out. Using modern carbon fiber and epoxy materials, it is possible to produce flywheels with rim speeds in excess of one kilometer per second and, depending on size, holding kinetic energy of several kilowatt hours. Under these conditions, flywheels are efficient and economical energy storage devices. Electrical transmission companies have a great deal of capital tied up in machinery that can be much under utilized during off peak periods. The possibility of taking advantage of slack periods to store energy in flywheels, which could then be employed for transmission during peak time demand, is a very attractive and highly efficient use of the available infrastructure.

A large magnet stores considerable energy. In consequence, another approach by which power utilities might accommodate cyclic changes in demand of electrical power would be to take advantage of such energy storage by superconducting magnets: one of the many applications of high field magnets. Let us now turn to look at these uses.

High Field Magnet Applications

In fact, there is increasing use in a wide range of high magnetic field applications of superconductivity:

- Laboratory magnets for research purposes,
- Magnetic energy storage devices,
- Magnetic resonance imaging for medical diagnostics,
- Superconducting magnets in nuclear physics: particle accelerators and bubble chambers,
- Magnetic levitated transport systems,
- Fusion and magneto-hydrodynamic power systems,
- Magnetic separation of minerals.

Of these enabling technologies for the 21st century, so far, only the first four items on the list have progressed beyond the prototype stage.

Magnetic Resonance Imaging

One of the most important and lucrative applications of high field superconducting magnets is in the medical field, especially in magnetic resonance imaging (MRI) now a major tool in medical diagnostics.

The discovery of x-rays in 1895 by Röntgen rapidly led to their use in the medical field through the ubiquitous x-ray picture. However, it was soon appreciated that they have three major drawbacks. Firstly, the x-ray picture is a two dimensional representation of a three dimensional object. All information relating to the depth of an object is lost and that coming from different depths can give rise to uncertainty in interpretation. Second, the intensity of a normal x-ray picture depends upon the amount of x-ray beam passing through the subject, which is determined by the attenuation from the different parts of the body; it is particularly good at revealing skeletal bones, which absorb much more strongly than tissue. X-rays are not good for distinguishing between different soft tissues with similar absorption, such as the liver and the pancreas. A third disadvantage with conventional x-rays is that since they record the average absorption due to all the different tissues through which they have passed, they are unable to distinguish between the separate densities of the individual tissues and therefore cannot provide quantitative information.

Starting from the early 1970s, a powerful new method for examining the human body using x-rays was developed which relied, in part, for its success on the rapid advances taking place in computer analysis. The pioneer was Godfrey Hounsfield, leader of a team based at the EMI Laboratories in London; the work led to his award of the 1979 Nobel Prize in Physiology or Medicine. Computerized tomography (CT) or computer-assisted (CAT) scanning measures the attenuation of x-ray beams passing through sections of the body at hundreds of different angles. Computer analysis of the raw data enables construction of pictures of the body's interior in three dimensions. A further decisive advantage is that the technique can be made enormously sensitive – allowing various soft tissues to be clearly differentiated. CAT scanning was a major breakthrough and also spawned other forms of tomography. Of these, nuclear magnetic resonance imaging appears to be the most successful having the advantages of CT but without using harmful ionising radiations. The last two decades have seen enormous development of this technique.

The technique of nuclear magnetic resonance (NMR) was discovered independently in the 1940s by Felix Bloch at Stanford and Edward Purcell at Harvard, and led to their joint award of the Nobel Prize for Physics in 1952. NMR has become a powerful tool for research in physics, chemistry and biology. Many of the techniques and much of the expertise gained in these areas has been invaluable for its extension into the clinical field. The name magnetic resonance imaging (MRI) became adopted – avoiding the emotive word "nuclear". The technology is relatively simple in that the basic requirements are a magnet, a radio antenna and a radio receiver. These days, superconducting magnets are used almost without exception. The nucleus of interest in MRI is the hydrogen proton held in water molecules within the body. When placed in a magnetic field, these hydrogen protons precess around the field direction with a definite frequency (called the Larmor frequency), which is proportional to the magnetic field intensity. In conventional MRI, a strong magnetic field is applied along the object under investigation so that the nuclei will align with their magnetic moments along this field. A signal at radio frequency, which is tuned to the precessional frequency of the hydrogen nucleus, is applied at right angles to the magnetic field by using a set of coils at the sides of the object. Some of the hydrogen nuclei in the body precess together in phase. When the radio frequency is switched off, the nuclei continue to precess in phase and generate a similar radio frequency, which can be picked up by a radio receiver placed close by. These precessing hydrogen protons detect the water content of the body. After some time, the precession dies away as the nuclei realign themselves once more with the magnetic field. The time that this takes, known as the spin-lattice relaxation time, is measured and provides the required information about the nature of the tissue under investigation.

The revolution in medical diagnostics using magnetic resonance imaging of previously invisible parts of the human body was kick-started by Paul Lauterbur at the State University of New York at Stony Brook. It was he who had the idea of employing a graded magnetic field which increased in strength across a body to provide a two dimensional image of the object's internal structure. Subsequently Peter Mansfield of the University of Nottingham showed that it is possible to image selectively a body held in a graded magnetic field by subjecting it to a blast of single frequency radio waves; he then developed the techniques required to piece images together in a matter of seconds. His work has enabled clinicians to follow fast moving events such as the beating of a heart. One of the most important clinical applications of MRI is in cancer diagnosis. Paul Lauterbur and Peter Mansfield have now been jointly awarded the 2003 Nobel Prize in Physiology or Medicine. Paul Lauterbur described his ideas in a seminal paper submitted to, and initially rejected by, the journal *Nature* in 1973. Fortunately his protests rectified the problem. Hours after the Nobel Prize announcement 30 years later, he told *Nature*, "I congratulate you on eventually giving in to my anguish!"

To obtain a scan, small sections across the body must be independently examined. In MRI, this is achieved by applying a small magnetic field gradient across the body, in addition to the main uniform field. The nuclei resonate at different frequencies across the body depending on the magnitude of the field gradient. In one method of MRI imaging, the frequencies received in the coil are separated from each other (using the well known mathematical method of Fourier analysis); the complete spectrum of frequencies represents a series of line integrals across the body and each frequency represents the number of hydrogen nuclei resonating along that particular line.

Early MRI systems used conventional permanent magnets. However, it was appreciated that superconducting magnets would have decisive advantages over permanent magnets by easily providing higher field strengths (between 1 and 2T), which are uniform over a diameter of about 1m. In addition, they produce extremely stable fields over a long time interval and can be made very reliable. The first super-conducting MRI system was manufactured by the Oxford Instruments Ltd and delivered to Hammersmith Hospital, London in 1979. Since then, over twenty thousand have been installed throughout the world and MRI has become the first large scale application of superconductivity.

Superconducting Magnets in Nuclear Physics

Superconducting magnets play an important role in providing magnetic fields in a number of machines used in nuclear physics, especially bubble chambers and particle accelerators.

A major advance in detecting high energy particles took place in the early 1950s with the invention of the bubble chamber by Donald Glaser; in consequence he was awarded the Nobel Prize for Physics in 1960. Glaser discovered that boiling in a superheated liquid could be triggered by the passage of cosmic rays or some other radiation. Incipient boiling makes the particle tracks visible. In the presence of a magnetic field, a charged particle can be identified from its curved trajectory in the bubble chamber.

A major goal of particle physics has been to probe deeper into the fundamental nature of matter by the brute force approach of colliding particles head on at higher and higher energies and observing the pieces produced. As the energy required to observe new particles becomes greater, more powerful magnets and larger bubble chambers are needed: superconducting magnets come into their own. There are advantages in that both these magnets and the bubble chambers require cryogenic temperatures to operate: integrated systems have become viable, culminating in the big European bubble chamber at CERN.

Great advances have occurred in elementary particle physics over the last few decades due in large measure to the building of bigger and bigger accelerators capable of producing particle projectiles of ever increasing energies. Two of the most prominent "atom smashers" are those at CERN in Geneva and Fermilab in Chicago, which are giant facilities which run on enormous budgets. Nowadays, almost all accelerators make extensive use of superconducting magnets and coils for bending and focusing the particle beams. The high and very stable fields which are available from superconducting magnets are a decisive advantage. Cost savings are significant, since superconducting magnets enable higher energy machines for a much lower electric power input; this more than offsets the expense incurred from the cryogenic cooling.

The insight gained from the vast quantity of experimental data generated by the various accelerators around the world has given rise to the "Standard Model" which provides a unifying description of the strong, weak and electromagnetic interactions. Theorists argue that this model is incomplete and only provides a partial explanation of the basic nature of matter. One worry is that it requires some seventeen "fundamental" particles and as many fitting parameters; it lacks the simplicity and elegance, which many physicists regard as an essential requirement of any fundamental theory. What seems to be needed is something deeper – grandiloquently termed a "Theory of Everything".

To provide experimental data to test such a theory, it is necessary to study particles colliding at even higher energies than those available so far – well into the TeV (i.e. 10^{12}eV) range. To achieve these high energies requires building and operating even bigger and more costly "atom smashers". The most ambitious accelerator proposed to date was the Superconducting Super Collider (SSC). This was designed to have two enormous magnet rings, each one 60 miles in circumference, that would accelerate protons to energy of 20TeV in opposite directions before making them collide. The circular orbits for the SSC were achieved by using superconducting magnets capable of reaching fields of 6.6T. As a result of having particles collide with the enormous energy of 40TeV, it was hoped that research using the SSC would generate a great deal of physics, which would shed light on ways to improve the Standard Model. In particular, it was hoped to observe the elusive Higgs particle first predicted in the mid 1960s. If that could be done, it would be an important step towards understanding the unification of the basic forces, as well as simulating some of the conditions in the very early Universe.

Because of the enormous cost of funding the SSC, the go-ahead for building it stirred up controversy right from the outset. It was argued that money would be better spent on more useful projects. Typical is a thoughtful and damning letter written in 1993 to the journal *Science* by Ted Geballe and John Rowell, two of the pioneers of applied superconductivity. The SSC project was cancelled in 1994 amid considerable acrimony in certain quarters of the scientific community.

No such fate has befallen the Large Hadron Collider (LHC) at CERN in Geneva. Already several years in construction, it is scheduled to come on stream in 2007. The LHC is a 27km circumference ring, which will produce protons colliding with each other with net energy of 14TeV. This is much greater than that thought necessary to detect the existence of the Higgs particle and also to explore its properties.

Management of the beams requires in excess of two thousand superconducting magnets of various geometries with state-of-the-art design. This project is the world's largest use of superconducting magnets. The experimental chambers form large cryogenic units that in operation are subject to unusually high heat loads requiring the development of complex superfluid helium heat exchangers. The development of the magnets is a truly international operation. In addition to that made by the twenty countries, which make up the membership of CERN, there are important contributions from Canada, India, Japan, the former Soviet Union and the U.S.

Superconducting magnet technology has enabled accelerators to attain higher and higher energies and beam flux. At the moment, the majority of the magnets operate at low temperatures using type II materials, although their replacement with high temperature superconductors is under consideration for future machines. The research stemming from the LHC accelerator and others currently under construction will increase our understanding of the fundamental particles of nature and conditions in the early Universe.

MAGLEVS

One of the most exotic, potentially exciting, and far reaching applications of high magnetic fields is in magnetically levitating transport systems or MAGLEVS. Because of the effects of friction and stability against vibrations, conventional wheeled rail transport is perhaps unlikely to exceed about 350km/hr, a speed which is now close to being reached by several railways but especially by the TGVs *(Train à Grande Vitesse)* of the French Railways. It is well known that a conductor moving through a magnetic field experiences a force of repulsion (Lenz's law). A similar repulsive force is experienced by a moving magnet passing over a conductor. In an actual MAGLEV system, there is an electrodynamic interaction between superconducting magnets placed on board a vehicle and coils on the ground causing the vehicle to levitate above a guideway, which corresponds to the rail track of a conventional railway. Prototype trains built on this principle have been developed and speeds in excess of 500km/hr have been obtained. The MAGLEV has some very attractive features such as higher velocity, lower pollution and reduced noise level compared with more conventional trains. However, it is a new and largely untested technology requiring cryogenically cooled, superconducting magnets. Unknown hazards might result from a failure of the superconducting magnets and there are fears as to whether MAGLEVS can be made robust enough to operate within a commercial environment. Pioneering work has been carried out in the U.S., Japan and Germany with heavy financial investment.

The Japan Railway Technical Institute has constructed a 43km test track in Yamanashi Prefecture some 100km west of Tokyo on the direct route through the mountains linking the major cities of Tokyo and Osaka. High speed MAGLEV trials have taken place, testing vehicles passing at speed through tunnels and especially entering and exiting, where severe turbulence can be created. If all the trials prove successful, the Central Japan Railway Company may decide to complete a commercial MAGLEV line linking Tokyo and Osaka and this would supplement the Bullet trains or *Shinkansen*, which currently join the two cities via the longer coastal route.

In Germany, tests on MAGLEVS have been underway for some 30 years starting with pioneering work at Siemens. Financial and technical planning is in advanced stages for building a prestige Transrapid MAGLEV system which would link Berlin and Hamburg. This project would do much to improve the east–west link within Germany.

Whether or not MAGLEV systems become viable forms of transport will depend on overcoming the numerous technical problems, as well as demonstrating the safety and reliability of such a mode of transport in competition with more conventional approaches. In addition, there are crucial economic, environmental and social factors, which have to be taken into consideration. However, the prognosis looks good. The world's first MAGLEV train system, operating between Shanghai's business center and its international airport, has successfully undergone trials and came into commercial operation in 2004. The trains will travel at speeds of up to 430 kilometers per hour and will cover the 30 kilometers route in about eight minutes compared with some 45 minutes by car. However, this particular MAGLEV system, which is a joint Chinese–German venture, uses conventional electromagnets rather than superconducting magnets. This illustrates that although prototype MAGLEVS exist using superconducting magnets, they will only enter into commercial operation if it is found that they are superior to those based on more conventional technology.

Magnetic Confinement of Plasmas in Fusion Reactors

Just as for years superconductivity has held out tantalising possibilities for widespread applications, so fusion has raised the prospect of almost limitless cheap energy. However, from the outset it has been recognized that the problems involved in the generation of electricity by a fusion reactor are formidable and are unlikely to be overcome until well into the present century or even the next. Fusion can only take place at extreme temperatures of several hundred million degrees. At such temperatures, the outer orbital electrons of all but the heaviest atoms are removed. The resultant collection of negatively charged electrons and positively charged ions form what is known as a plasma, characterized by having a high electrical and thermal conductivity. The plasma state is regarded as an important and intriguing fourth state of matter, after solids, liquids and gases. A plasma has to be confined within a container without touching the sides, which would be immediately vaporized by it. Because a plasma is a good conductor of electricity, it can be confined or trapped within a suitably shaped "magnetic bottle" for a long enough time for the fusion process to take place. Such a bottle is usually formed by a clever arrangement of strong magnets. The principle is that the ions and electrons can no longer travel in straight lines in the magnetic fields but instead move in helical orbits; this impedes their escape. Confinement is achieved using superconducting magnets since only they can provide both the necessary high fields of several Teslas and the required field gradients. The simplest form of confinement is in a ring or torus, which is the basis of the best known configuration: the "Tokamak". At the moment the magnets to keep the plasma circulating within a ring are almost all made from type II materials. Eventually these materials are likely to be replaced by high T_c super-conductors because of the reduction in both the cost and the complexity of the cryogenics. However, this will only occur when magnets can be made from high T_c materials which are good enough to operate reliably under stringent conditions of high fields and current densities.

The world demand for energy is increasing at an alarming rate. One third of the world's exploding population remains without electricity. It is believed that there may be sufficient fossil fuel for at most two hundred more years. But burning fossil fuel has the drastic disadvantages of creating the carbon dioxide burden in the atmosphere, contributing to the greenhouse effect and helping to accelerate global warming; in addition it presents health and safety problems. At one time, it appeared that nuclear power was the answer to the world's energy problems. However, distrust, much exacerbated by the Chernobyl and Three Mile Island disasters, has ensured that it is unlikely that nuclear power will become anything like as widespread as was once thought.

It has always been recognized that the fission process responsible for nuclear power produces radioactive by-products and the safe handling and storage of these poses a formidable problem, which will continue for hundreds of years. More benign sources of energy such as hydroelectric, wind, wave and solar power have attractive environmental properties but are considered to be unlikely to produce more than a few percent of the total world energy requirements. Compared with fossil fuel and nuclear power, fusion has enormous potential for an environmentally friendly and cheap source of almost limitless energy. In fusion, light nuclei bond together to produce heavier nuclei but the end product has slightly less mass than the original components. Like fission, fusion relies on that most famous of all physics equations: $E = mc^2$. What Einstein's equation means is that if a nuclear reaction occurs in such a way that there is less mass present in the end products than there was at the beginning, then the lost mass (m) will be converted into energy (E). Because the velocity of light (c equal to $3 \times 10^8 \mathrm{ms}^{-1}$) is so large, and its square is involved, conversion of a small amount of mass creates a huge amount of energy. That is why so much effort is being put into making fusion reactors a reality. A decisive advantage of fusion over fission is that the end products are neither radioactive nor in any way hazardous.

The problem is that fusion can only occur under extreme conditions. As long ago as the early 1930s, it was appreciated that a fusion reaction, converting hydrogen to helium, is the source of the energy from the sun. Similar conditions to those existing in the interior of the sun have to be reproduced in the laboratory to obtain fusion here on earth. A promising reaction for energy production is the fusion of the heavy hydrogen isotopes deuterium and tritium, which gives rise to harmless products and a large release of energy. The heat energy liberated is used to produce the steam, which drives the turbines to generate electricity in a conventional manner. Both deuterium and tritium are comparatively easy to produce in almost limitless amounts raising the distant prospect of cheap and abundant electrical energy.

Magnetic Separation of Minerals

Superconducting magnets are already operating in a truly industrial environment in magnetic separation of minerals from ores. In Chapter 2 it was mentioned that weakly magnetic materials can be classified as either paramagnetic or diamagnetic depending on whether they are attracted or repelled by an external magnetic field. Magnetic separators operate upon the principle that the force on a mineral particle is proportional to the magnitude of an applied field and its gradient. Materials to be separated are suspended in a liquid to form a slurry and this is passed through a magnetic field. Different particles are deflected by different amounts, resulting in separation.

To date, the main application of this process is in the purification of kaolin or china clay, which has widespread applications in the ceramic and paper-making industries. Kaolin is nonmagnetic whereas many of the unwanted colored impurities like iron oxide are paramagnetic. Removal of impurities improves the kaolin by making it whiter. Superconducting magnets are able to achieve much larger magnetic fields and field gradients than electromagnets as well as consuming far less energy. Most of the superconducting magnets used depend upon liquid helium for cryogenic cooling. Some mines producing kaolin, such as those in places in South America, are very inaccessible; transporting liquid helium from the point of production to the site of the mine requires a long journey resulting in the loss of a substantial fraction of the cryogenic fluid. A high temperature superconducting technology based on the use of liquid nitrogen would be enormously preferable provided the magnets are reliable and can reach sufficient current densities.

The Future?

The exploration of large-scale applications of superconductivity is on the increase. However, the more "Gee-whiz" and zany suggestions, like solving traffic congestion by people using superconducting skis, have yet to get off the drawing board and remain science fiction.

Nevertheless technological progress made since the discovery of the high temperature superconductors is impressive. The use of superconducting magnets in the two very different areas of fusion and magnetic separation anticipates a changed world, in which fossil fuels for energy production have been largely depleted and raw materials have become scarce, making recycling essential. A range of technologies using superconductors will become feasible and commercially viable as this century proceeds.

It was the genius and vision of Kamerlingh Onnes to recognize that super-conductivity had technical possibilities for distributing electricity and producing high magnetic fields. It was he who also first appreciated the fundamental limitations of the superconductors available to him. Overcoming the problems that he recognized has taken a long time but as the centenary of his discovery looms up the prospects that superconductivity will form the basis for enabling technologies in the twenty first century look good.

9 Electronic Applications of High Temperature Superconductors

In the euphoria following the discovery of high temperature superconductors, many extravagant claims were made of bright prospects for a new industrial revolution. By the time that reality had taken over from the excessive hype, a jaundiced view of the future of superconducting technology prevailed. Things had gone from one extreme to the other.

Today a wide range of applications of the new high temperature superconductors in both industry and research has come on stream: telecommunications, systems used in space technology, high speed computing, ultra-sensitive measuring instruments. Initial commercial applications have been in electronics. This is because the technological problems involved in fabricating small devices are less difficult to overcome than those encountered during the manufacture of the wires and tapes required for large scale applications (Chapter 8). However this success has only been achieved following great efforts to overcome the formidable problems of making the high quality thin films needed to make electronics devices. It is a tribute to the ingenuity and determination of those involved that such films were successfully produced within a few years of the initial discovery of these ceramic materials.

In attempting to grow thin films of high temperature superconductors, scientists have been able to draw on the experience obtained from conventional superconductors and other materials. Good quality thin films of the superconducting elements lead and niobium are fairly easy to produce and were much used in early applications. Far greater problems were encountered and overcome in producing thin films of niobium nitride and even more of the really useful A-15 materials such as Nb_3Sn.

The considerable commercial spin off from using thin films of the high temperature superconductors has justified the great effort that has been put into producing them. Most of the films have been made of YBCO, although there has also been quite extensive work on the bismuth, thallium and mercury based cuprate superconductors (Chapter 4). To make films, it has been necessary to turn to highly sophisticated, state of the art techniques including co-evaporation, molecular beam epitaxy (MBE), sputtering, pulsed laser deposition, and metal organic chemical vapor deposition (MOCVD) (these are described in Box 11). To produce devices for microwave and other applications, the thin film of the superconductor is deposited onto a suitable substrate (Box 12). The manufacture and monitoring of good quality

Box 11

Production of thin films of high temperature superconductors for devices

The properties of the highly anisotropic, layered cuprate compounds, which contain three to five metallic elements as well as oxygen, are very sensitive to crystal structure and oxygen content: problems not encountered with the low temperature elemental superconductors. What is more, the constituent elements of the cuprates are extremely reactive, and this limits the choice of suitable substrates. Further complications arise because films can only be laid down at high temperatures, in the range 700 to 900°C, and there are difficulties in ensuring that the required superconducting phase is produced rather than one of the other stable phases. All these factors highlight the technical problems encountered in producing high quality thin films of the high temperature superconductors.

Standard methods of thin film production
1. The first technique used to grow films of high temperature superconductors was **evaporation.** YBCO films can be made using independent sources of the three metal ions, yttrium, barium and copper (Figure 9.1). Separate sources are

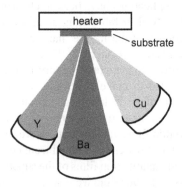

Figure 9.1   Method employed to make films of high temperature superconductors, such as YBCO, by evaporation from separate sources of the three cations, yttrium, barium and copper. The vapors are produced by heating the sources or directing an electron beam onto them. The flux from each source is carefully controlled to ensure that the film deposited onto the substrate has the required composition ratio.

necessary because the vapor pressures of these components differ. The vapors are produced either by heating the sources or by directing an electron beam onto them. The flux from each source has to be carefully controlled to ensure that the film deposited onto the substrate has the required composition ratio: Y: 1, Ba: 2, Cu: 3. Control is assisted by monitoring ion flux using optical absorption. Sufficient oxygen has to be available to ensure that the superconducting oxide

$YBa_2Cu_3O_{7-x}$ contains the correct oxygen concentration $(7-x)$ (see Chapter 4). Film growth is achieved either by simultaneous evaporation of all the components or alternatively by shutting off each component sequentially.

2. Molecular beam epitaxy (MBE) is a sophisticated advance on simple vacuum evaporation much used to make high quality films layer by layer. This technique, developed by the semiconductor industry and invaluable for fabricating devices, now finds further use in making high temperature superconducting films. An epitaxial growth process is one in which the arrangement of the atoms of the newly added material is a continuation of the ordered crystal structure of the "substrate" onto which it is grown. During MBE growth the components are directed in a beam onto the substrate surface inside a very high vacuum chamber (Figure 9.2). Each component source is heated in a separate "effusion cell" in which the source temperature determines the number of molecules in the beam.

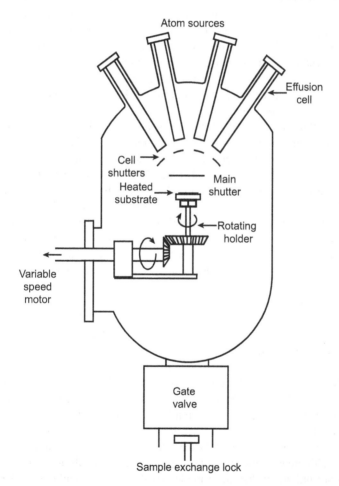

Figure 9.2 A typical molecular beam epitaxy (MBE) machine. The thin film is deposited layer by layer onto a heated rotating substrate in a high vacuum chamber.

To select a beam of each different species rapidly in turn, an externally controlled shutter is placed between the outlet of each effusion cell and the substrate. A controlled beam of molecular vapor is allowed to escape from the effusion cell through an orifice that is much smaller than the distance that the particles travel between collisions with each other in the beam. Ion gauges precisely monitor the intensity of each molecular beam. This enables control of the chemical composition and the growth rate of the epitaxial layers. The substrate is rotated to ensure even layer growth over its surface and heated to allow the atoms falling on it to move into the positions needed to form a satisfactory film structure. A major advantage of the MBE growth technique is that the sample can be monitored whilst growth is taking place: many MBE machines are equipped with sophisticated electron beam apparatus that can characterize the surface of the material – such as reflection high-energy electron diffraction (RHEED). Electrons are chosen to make this diffraction examination because they can be excited to a wavelength smaller than the interatomic spacing of the epitaxial material. Remarkably, it is now possible to control the growth of the film to a precision of one atomic layer. Nearly perfect layer by layer material can be made. Since the high temperature superconductors are layered materials, MBE is a particularly suitable technique to produce films. In addition it is ideal for the production of novel multi-layered structures and therefore makes possible the search for new high temperature superconductors.

3. Sputtering is another technique used extensively by materials scientists for depositing thin films. In this process, atoms are ejected from a cathode target by bombardment with energetic ions produced in a gas plasma. These atoms are then deposited onto a nearby substrate. Maximum deposition rate can be achieved in the on-axis geometry in which the substrate and target face each other. However, this configuration suffers from the severe drawback that the deposited film may be damaged by bombardment with ions from the gas plasma. To avoid this, the substrate surface is positioned at right angles to the sputter target. This off-axis configuration has the severe disadvantage that the yield is reduced considerably. Another drawback of the sputtering technique is that the film deposited may not have the same composition as that of the multi-component target due to differences in the sputtering yields for different elements. In spite of these problems this inexpensive technique continues to be heavily used for making high temperature superconducting films.

4. Pulsed laser deposition has also come into prominence as a result of its proven success in producing good quality films of the high temperature superconductors. In this method, a pulsed laser is focused onto a target of the material to be deposited. Provided the laser has a sufficiently high energy density, each laser pulse vaporizes, or ablates, a small amount of the target material. A plume of vapor, ejected in a forward direction, provides the material flux for film growth. This technique is much faster than off-axis sputtering and is well suited to the stoichiometric deposition of complex oxide materials and multi-layers. Its widespread utilization has been made possible by the availability of reliable, high-powered lasers.

5. In the metal-organic chemical vapor deposition (MOCVD) method the cations, which are required for film growth, are delivered to a heated substrate as the constituents of organometallic molecular vapors. The hot substrate causes the source compounds to decompose at the surface so that epitaxial growth can proceed. This technique is extensively utilized in the semiconductor industry because it allows film deposition over a large surface area at a high throughput. It has proved relatively easy to adapt MOCVD technology to produce films of high temperature superconductors. The main problems are associated with controlling the rates of arrival of the different elements needed to form the superconducting film. However, MOCVD is being used on an increasing scale and is likely to become even more important in the future.

In the case of YBCO, the most widely used high temperature superconductor in electronic applications, the different deposition techniques described tend to produce films of similar quality. In each case, the films can be made homogeneous and can pass current densities in excess of a critical value of 10^6 amperes per cm^2 at liquid nitrogen temperatures (77K) and zero applied field. The critical current densities fall off slowly with applied fields but values in excess of 10^5 amperes per cm^2 in magnetic fields as large as 20 Teslas have been achieved. Work is always in progress to improve film quality and performance.

superconducting films suitable for commercial applications remains an active area; innovations are continually taking place.

The use of superconducting films in electronics can be broadly divided into two categories: passive and active devices. Passive devices generally modify a single electrical signal or input. Active devices tend to have another signal controlling the modification of the input in a non-linear manner. Passive devices are often used in microwave applications, while Josephson junctions are examples of active devices.

Box 12

Choice of substrate for film deposition

For use in high temperature superconducting electronics, large area (diameter typically of about three inches) substrate surfaces can be needed. Any imperfections in the film can degrade device performance. An essential requirement is that the crystal structure and spacing between the atoms of the substrate and those of the superconductor should be compatible; otherwise strains would be set up in the films giving rise to defects and dislocations. The substrate is usually obtained from a bulk crystal, which is cleaved, polished and then etched chemically to provide a clean, atomically flat surface free from steps, islands or unwanted impurities since these can give rise to imperfections in the growing film. Ideally, a substrate should provide good mechanical support for the thin film but not interact chemically with it in any way: it should be environmentally stable. Absence of chemical interaction is especially difficult to achieve for the high temperature superconductors since their component elements are rather reactive. It is usually necessary to grow the film at a relatively high temperature, in excess of 700°C, at which it is difficult to keep a substrate chemically inert. Choice of the substrate onto which the thin film is to be deposited depends on the superconductor and also on the application in mind. The best YBCO films have been put down onto perovskite substrates such as a single crystal of strontium titanate ($SrTiO_3$) or lanthanum aluminate ($LaAlO_3$). These insulating perovskite compounds are related structurally to the cuprates and have in addition the benefit of possessing similar lattice spacings to those in the super-conducting a–b layer plane of YBCO (and other layer-like cuprates) (see Chapter 4). As a result, it is possible to deposit good strain-free superconducting layers on these crystal substrates: just what is needed to make devices. They have been used extensively, although they suffer from undesirable dielectric properties that reduce their suitability for very high frequency applications. Non-perovskite substrates have also been much used. For example, excellent superconducting films have been deposited on magnesium oxide (MgO) substrate but unfortunately this material can suffer from poor mechanical properties as well as tending to absorb water. Artificial sapphire (monocrystal-line Al_2O_3) and single crystal yttria–stabilized-zirconia (acronym YSZ) have also been used successfully as substrates on which to grow YBCO films.

Buffer layers
Another approach has been to find suitable buffer layers. These are layers that are grown directly onto the substrate in order to reduce the effect of some of its shortcomings. Some of the problems arising from lattice and thermal expansion mismatch can be reduced by using a suitable buffer that has a lattice spacing intermediate between that of the substrate and the superconductor. Buffer layers can also minimize inter-diffusion and chemical reactions between the super-conducting film and the substrate and may well be crucial in obtaining the best match for future high temperature superconductor applications.

To take full advantage of the high temperate at which superconductivity occurs in the cuprates, it is important to design low loss devices. Since some of the commonly used substrates such as $SrTiO_3$ and $LaAlO_3$ have large dielectric losses, considerable effort is currently being expended to develop suitable substrates and buffer layers, which meet the desired requirements.

What are Microwaves?

Arguably the most important piece of theoretical physics in the nineteenth century was the development of electromagnetic theory by James Clerk Maxwell, who was a minor Scottish laird. In his work, Maxwell showed that light is comprised of electric and magnetic fields, oscillating sinusoidally at right angles to each other. A new and fundamental aspect of his electromagnetic theory was its implication that visible light formed only a small part of a large electromagnetic spectrum, which it suggested extended to both higher and lower wavelengths and frequencies.

There was already some evidence for the existence of electromagnetic waves outside the visible range of the human eye. One of the great scientists of the late eighteenth and early nineteenth century was William Herschel. He is particularly revered in Bath since it was from a garden in this city that in 1781, he discovered the planet Uranus, the first planet to be discovered since antiquity. In 1800, long before Maxwell's work, William Herschel had examined the effectiveness of different regions of the optical spectrum in raising the temperature of a thermometer. He discovered that the greatest temperature rise occurred in a region beyond the red end that became known as the infrared. The following year the German scientist Johann Wilhelm Ritter discovered that an invisible radiation beyond the violet end of the spectrum darkened silver chloride more effectively than visible light; this spectral region became known as the ultra-violet. Maxwell died prematurely from cancer aged 48 in 1879 (the year that Einstein was born) sadly before his ideas about the electromagnetic spectrum had been fully tested. Verification came about in 1888 through the brilliant researches of the young German scientist Heinrich Hertz, who was able to generate and detect radio waves and also to show that, in a similar manner to light, radio waves could be reflected, refracted, diffracted and polarized. Most importantly he established that the velocity of radio waves was exactly the same as that of light – as predicted by Maxwell. German physics was dealt a severe blow when in 1894 Hertz died young at 36 from blood poisoning allegedly due to bacterial contamination in his laboratory, which had formerly been used as a mortuary.

The electromagnetic wave spectrum continues beyond the infrared into the microwave region. Microwaves are radio waves with very short wavelengths (between about 0.3mm and 30cm), as the prefix *micro* implies. Officially, the microwave range begins at a frequency of about 1000 megahertz (10^9 cycles per second), above that of ultrahigh frequency radio waves (UHF), and goes up to about 300 gigahertz (3×10^{11} cycles per second). Microwave wavelengths are about the same size as ordinary electronic components: devices operating at microwave wavelengths are small enough to put into aircraft for radar navigation.

During the Second World War, the need for detection of enemy airplanes and ships at a distance gave impetus to the development of radar: location by the use of radio echoes at microwave frequency. Knowing that the velocity of microwaves is the same as that of light, it was possible to detect and time echoes reflected from a target, which is much larger than the wavelength of the microwave signal. Devices to produce and receive microwaves were needed because the vacuum tubes in use in the 1930s did not operate at the high frequencies required. A microwave receiver had already been available since 1937, when Russell and Sigurd Varian had invented the klystron, a vacuum tube that amplified microwaves. The first practical high power microwave generator was the magnetron, which remains in use today in some radars (and in microwave ovens). Magnetrons were soon made small and light enough to be carried in airplanes. Airborne radar was crucial for the Allies to win the war against submarines in the North Atlantic.

Today, electronic engineers use a wide variety of solid state devices in radar, microwave communications, electronic navigation and satellite TV links. Recently high temperature superconducting devices have established a position in many applications at microwave frequencies, including mobile (cell) phone central systems. Developments have been rapid: superconducting microwave devices can now be bought in the marketplace and communications packages are being flown in space.

Applications of High Temperature Superconductors in Microwave Technology

The best known property of a superconductor is that it shows no resistance to current flow. Both superelectrons and normal ones are present in a superconductor, a feature expressed in the two fluid model (see Chapter 2). The superelectrons carry all the current. Actually, this is only strictly true for a direct current, which of course keeps a constant value. Electrons have an inertial mass. So when an alternating voltage is applied, the electrons can't keep up with it and the supercurrent lags behind the field. This means that there is an inductive impedance in a superconductor carrying an alternating current; due to the time lag, an electric field is present and some of the current is carried by the normal electrons which are present in addition to the superelectrons. Since these normal electrons are scattered by thermal vibrations and defects in the usual way (see Chapter 5), there is some electrical resistance. As a result, so far as an alternating current is concerned, a superconductor is best represented as a perfect inductance to the superelectrons in parallel with a resistance to the normal electrons. Microwaves produce an alternating current in a superconductor so their absorption is also determined by the presence of both normal and superelectrons. As the frequency is increased, the impedance to current flow, due to the inductance, becomes larger. When the frequency of visible light is reached, the impedance has become so large that no difference can be seen in the appearance of a superconductor and its normal counterpart: the reflection of light from the surface is essentially the same for both phases. The fact that when a metal goes superconducting it does not change in physical appearance was observed long ago.

Ordinary electronic components do not operate at microwave frequency. One reason for this is that component sizes and lead lengths are of the same magnitude as the signal wavelength; another is that inductance and capacitance become significant at high frequency. A third is that alternating currents tend to flow on the surface of a metallic conductor. This "skin effect" results in a microwave current being confined to the "skin depth" in the normal state, while they are contained within the penetration depth (see Chapter 2) in the superconducting state. The major advantage of using superconducting films rather than ordinary metals in microwave devices is their much lower surface resistance. For an ideal BCS superconductor, the microwave losses are negligible at very low temperatures. In high temperature superconductors, the losses are much greater than that expected for the ideal BCS case and arise from several sources including surface imperfections and magnetic flux trapped at the surface. Nevertheless the surface resistance of these materials can still be very small: one to three orders of magnitude less than that of a normal metal such as copper at liquid nitrogen temperature 77K: superconductors are more efficient. Small, high

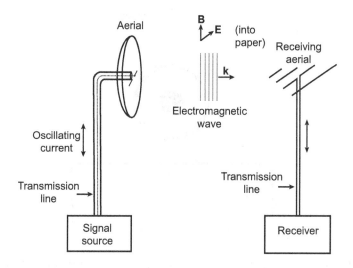

Figure 9.3 The general layout of a system for transmitting and receiving radio waves. Microwave systems have similar features but incorporate components designed to operate at the higher frequencies involved. When an electromagnetic wave is propagating through space, its associated magnetic **B** and electric **E** fields are at right angles to its propagation direction (along vector **k**).

performance, microwave devices are the preferred option at this temperature and applications of high temperature superconducting thin films in microwave devices abound. The two-dimensional character of devices made from films enables easy access – facilitating design and construction using modern lithographic production techniques. Examples of devices operating at microwave frequencies include: low noise oscillators, filter circuits, delay lines, antenna arrays and very high quality Q cavities. We need to see how these are made and work.

Before discussing their applications in microwave electronics, it is worth noting briefly that microwave experiments provide important information about the fundamental behavior of cuprate superconductors. Measurement of the microwave surface resistance has become a standard technique for determining the quality of a high temperature superconductor. At liquid nitrogen temperatures the best epitaxial thin films of YBCO outperform copper up to 500GHz. This situation is likely to improve, as film quality gets better.

To transmit radio waves, a current, which is suitably modulated with the information being carried, is sent along a transmission line to an aerial (more formally called the antenna) (Figure 9.3). The antenna transforms the oscillating currents into electromagnetic waves. These travel across space to a receiving antenna in which they induce currents that then propagate along another transmission line to a receiver. There the original information contained is reconstructed. Microwave systems also work in this manner but they must incorporate specific design features due to the higher frequency of microwaves. Superconductors play a useful role in the component parts of microwave systems designed for a number of specific applications.

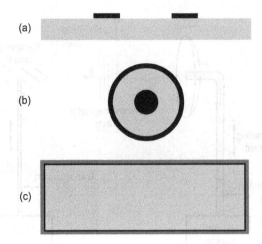

Figure 9.4 Cross sections of transmission lines. In each case the dielectric is shown shaded in light grey. (a) The simplest type is comprised of two parallel conductors; this can be made by depositing two strips of superconducting film (black) onto a strip of insulator (dielectric). (b) The best known is the coaxial cable used to transmit a television input signal. (c) The performance of a rectangular waveguide can be improved by an internal coating of super-conductor (black).

Superconducting Transmission Lines

Transmission lines are used to convey high frequency signals from one part of a system to another. Simple and commonly used types of line are shown in cross section in Figure 9.4. The oldest and archetypal form is made from two conductors placed parallel to each other and separated by an insulator; the modern use of printed circuitry has given such parallel lines a new lease of life. The conducting lines can be made by depositing two strips of superconducting films onto an insulating support strip (Figure 9.4(a)). The most widely used transmission line is the coaxial cable, which is made from two concentric metal cylinders separated by a dielectric (that is, an insulator such as a plastic), the cross section is shown in Figure 9.4(b). Coaxial cable is familiar to us all because it is used in our homes to connect the aerial to our television set.

Transmission lines become increasingly lossy as the signal frequency is increased. At higher microwave frequencies, these losses become unacceptable and waveguides are used. What is a waveguide? A useful analogy is an optical fiber used as a light pipe. The fiber surface acts as a mirror containing the light within the fiber: most of the light energy sent into the input end of the fiber is delivered at the output end. A waveguide is a metal tube, which behaves as a pipe for electromagnetic waves at radio frequency: it can have either circular or, much more commonly, rectangular cross-section, as shown in Figure 9.4(c). Most waveguides are made from an electrical conductor such as aluminium, brass or copper. Waveguides can now have their

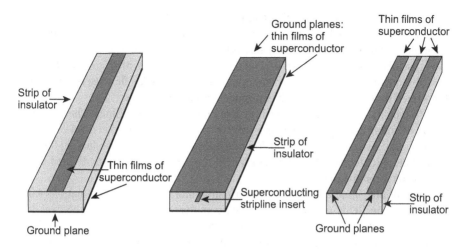

Figure 9.5 Types of superconducting waveguides. The conducting lines, including those to ground are made from superconducting films (shaded dark grey): (a) microstrip, (b) stripline (inserted along the center of the insulator), (c) coplanar. In each case, the dielectric is shown shaded in light grey.

internal faces coated with a high temperature superconductor to improve their performance by reducing the impedance to the microwave signal.

An important application of superconductors is in low loss transmission line. One of the simplest, but most useful, forms of transmission line at microwave frequencies, is microstrip, which consists of a strip of conductor (whose width can be critical) placed above a ground plane from which it is separated by an insulating layer of dielectric (Figure 9.5(a)). An advantage of superconducting microstrip is that the penetration depth and propagation velocity of the signal do not vary with frequency; hence very short pulses, which contain many frequency components, can travel relatively long distances down such a line without altering shape. Planar waveguides come in a number of different guises based on this general type of design, including those shown in Figure 9.5. One of these, stripline is made by using two ground planes having a gap between them and with a central signal line deposited along the middle of the gap (Figure 9.5(b)). A third widely used form is coplanar line (Figure 9.5(c)); this has the advantage that it can be made by deposition of superconducting film on one side only of the insulator and can be used for planar devices that can be integrated into conventional microwave circuitry. An increasingly important application of superconducting transmission lines is for interconnecting silicon chips in computers.

Superconducting Delay Lines

Superconductors are also used in delay lines; these consist of a long transmission line usually made in the form of a superconducting microstrip, stripline or coplanar line deposited onto an insulating substrate; that is one of the types shown in Figure 9.5. In general delay lines form an integral part of radar, computer and communications technology. The purpose of a delay line is to store a signal for a substantial period of time (usually less than a millisecond) and then to recover it with minimum distortion for subsequent use in display or computing operations. Delay devices retain a signal for a fixed time interval rather than provide permanent storage.

In fact, many techniques have been used for making delay lines. These include: transit of an ultrasonic pulse across a solid or along its surface as in a SAW (surface acoustic wave) device, a light flash down an optical fiber, a superconducting line. The delay time is determined by the path length traversed by the pulse divided by the pulse velocity in the line material.

Essentially, a delay line is an information storage device; applications range throughout the fields of communications, commerce and industry, wherever long delays and high fidelity are required. An early use was in moving-target indicator radar systems for aircraft detection. In these systems, the delay line stores the complete set of echoes that are reflected from the target from a single output pulse; then these are balanced against reflections received from successive pulses. Delay lines are widely used in radar, as well as in many other fields: as timing elements such as time delays, as range and time markers, as frequency and time-interval generators, in interferometers, in television and as part of pulse-forming or compensating networks in multichannel communication systems, in beam steering antenna systems, and many electronic warfare and satellite communication systems. A delay of a signal can be needed while the input signal undergoes fast reprocessing; in such a case, the time scale is a few hundred nanoseconds (several tenths of a millionth of a second) for which a superconducting transmission line can be well suited.

It is worth noting that ultrasonic delay lines are used particularly when long delay times coupled with high fidelity are needed. The velocity of ultrasound waves propagating in a solid is only about 10^5 cm per second; that is, a factor of 10^5 times slower than that of electromagnetic waves in a vacuum. Hence the transit times for acoustic waves are greater by a factor of 10^5 than those of electrical pulses: long delay times are available. Why can long delay times be necessary? A simple example provides the answer. A ten-digit number can be represented by 34 binary bits. If the clock-pulse repetition rate is 2MHz, then one ten-digit word uses 17 microseconds. In the case of a computer with a total storage capacity of 1,000 words, a delay line for a time of 17 milliseconds would suffice to supply all the storage. Ultrasonic delay lines can be made which store pulses as short as a tenth of a microsecond in duration and with delay times 4 milliseconds long, equivalent to storage of 40,000 bits of information.

In the case of superconducting delay lines, the propagating electromagnetic waves are 10^5 times faster than ultrasonic waves so the delay is shortened by that factor. These lines are easy to fabricate and have a number of advantages for certain applications. They are amongst the lowest loss devices used for providing a delay. Much emphasis has been placed on design of compact delay lines with a wide frequency bandwidth and low spurious response. High frequencies, usually in the microwave range, must be used for propagation of short pulses or the output will be distorted. In his book *Passive Microwave Device Applications of High Temperature*

Superconductors, Mike Lancaster has listed a number of superconducting delay lines made by several companies. To save space, the majority of these devices are made planar as a spiral of microstrip or stripline.

Superconducting Filters

Electrical transmission line systems, or delay lines, are essentially filters comprised of a series of inductances and condensers. A great advantage of superconducting delay lines is that they can use the delaying process to filter the input signal (or perform a number of other useful processing operations on it).

In general, a filter is a device or circuit that positively discriminates for some frequencies in a signal; these favored frequencies, which are allowed to proceed, are known as the passband. Alternatively a stopband filter can be made which selectively discriminates against selected frequencies. Band pass filters must have a sharp response curve so that they allow only the required signals to pass while rejecting others. Superconductors are used in the construction of certain filters since they can improve performance by eliminating resistance in the filter circuit; the bandwidth can also be reduced and the steepness of the cut-off at the skirt of the band can be sharpened. The quality factor of a filter made with conventional components is typically about 2500. That of the first generation superconducting filters is 40000 and values could well reach a million. Such a filter acts like a brick wall to unwanted signals.

Another advantage is miniaturization; superconductors enable a substantial reduction in size compared with filters made from normal metals, although such miniaturization requires new types of filters with different geometries. A wide-ranging, specialist account of superconducting filters has been given in the book by Mike Lancaster. Superconducting RF and microwave filters based on microstrip and cavity designs for wide-band communications and radar are now commercially available.

Currently there is rapid progress in the application of high temperature super-conducting filters in cellular telephone systems. The use of mobile telephones is increasing by about 30% per year and most localities have more than one cellular phone provider. These phones operate in the frequency range 800 to 900MHz and most cellular phone companies have several frequency allocations within this band. Filters are needed to ensure that signals from one cellular phone provider do not interfere with those of another. Filters made out of high temperature superconduct-ors, especially YBCO, are set to replace the conventional copper filters used in cellular base stations where it is easy to introduce a technology that operates at liquid nitrogen temperatures. These high T_c filters have decisive advantages: a sharper response and smaller losses than those made of copper (even when the copper is operated at 77K). Several companies in the United States, Europe, and Japan are actively developing this area of telecommunications, and it is widely felt that this will become the first major market for high temperature superconductors with potential sales in excess of $100 million per year.

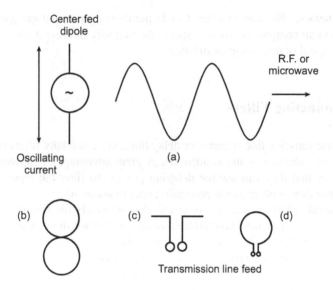

Figure 9.6   Dipole aerials: (a) operation, (b) figure of eight pattern of signal. Simple aerials can be made of superconductor and attached to superconducting transmission lines: (c) dipole, (d) loop antennas.

Superconducting Aerials

Aerials are used to transmit and receive radio waves from the lowest to the highest frequencies. Superconductors are used not only as transmission lines to convey a microwave signal to an antenna but also in the construction of the antennae themselves. Microwave technology makes use of both radio and optical techniques: dipoles and parabolic reflectors find specific niches as antennas (Figure 9.3). A dipole is a two pole antenna which can be made either as a single radiator fed at its center or as a pair of elements (Figure 9.6(a)). A basic requirement of an antenna is to match the impedance of the transmission line leading to it with that of free space; this optimizes the power transfer from the transmitter to free space (or alternatively from free space to the receiver). A microwave dipole often consists of a short piece of conductor placed at the end of a section of waveguide or a simple termination at the end of a superconducting microstrip as shown in Figure 9.6(c). Although a dipole should ideally be a half wavelength long, in practice, it may not be. For many applications, it is efficient to have the antenna about a single wavelength long and made of a normal metal. However, when it becomes necessary to reduce the antenna size, resistive losses increase and superconductors become useful. Once strong sintered YBCO wire became available, it was feasible to make simple dipole antennas, superior in performance to those made with copper.

The radiation patterns produced depend upon aerial design. Signal directivity is a fundamental property of an antenna: a dipole (Figure 9.6(a)) directs energy in a figure of eight pattern (Figure 9.6(b)). For some applications a different directionality

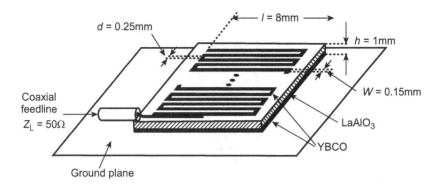

Figure 9.7 A meander line superconducting antenna used to produce sharp directionality of signal. (After H.J. Chaloupita (1992).)

may be needed. The commonly used loop antenna (Figure 9.6(d)) is made nowadays with YBCO wire.

Depositing appropriate patterns of thick YBCO films onto a substrate is also used to make antennas, which have orders of magnitude better performance than those made from copper. An advantage is that films can be deposited in complex but precisely defined structures, which produce a desired radiation pattern. An example is the directional meander line antenna for operation at a frequency of 940MHz shown in Figure 9.7. The resonant frequency of a meander antenna corresponds to the total length with the bends removed. Such antennas, fabricated by silk-screen printing, can have high radiation efficiencies even though they are small in size. Another advantage is that they can be made with a very sharp directivity as compared with those of the simple dipole shown in Figure 9.6(a): superconductors are enhancing the scope of highly directional antennas.

Superconducting Microwave Cavities

At frequencies lower than those of microwaves, tuned circuits, which are combinations of inductors and capacitors, are used to make oscillators and filters. When you tune in to your favourite program, you are changing the component values in the tuned circuit situated at the input end of your radio to make it resonate at the same frequency as the station transmits. However at higher frequencies, the inductors required become impractically small, and so, resonant cavities take over the role played by tuned circuits at lower frequencies. Customarily microwave cavities are metal walled, cylindrical, spherical or pill-box shaped and, like an organ pipe, can resonate in modes with wavelengths related to the cavity dimensions. There is a small aperture in one of the walls through which the signal can enter or leave.

Cavity resonators are a fundamental building block of a microwave circuit ensuring that it operates at the required frequency. For a changing current some power is dissipated. A basic parameter of a cavity is its quality factor Q: the ratio of

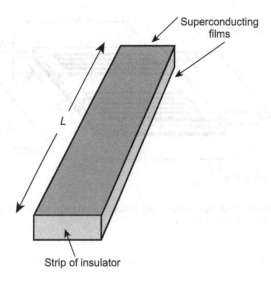

Figure 9.8 Microstrip cavity resonator. The length L of the cavity can contain an integral number of half wavelengths of microwave and so can resonate at these values. This is akin to the situation in a vibrating violin string.

the energy stored to the energy loss per cycle. Coating the walls of a cavity with a superconducting film, or forming them from a superconducting material, can reduce losses and so improve the quality factor. A thick film coating can be made cheaply by spreading an ink-like mixture of superconductor and a paste onto the surface to be covered and then heat curing. Such practices have become popular following the emergence of high temperature superconductors. Since the quality factor Q of a device is inversely proportional to the microwave surface resistance, it is possible to obtain extremely high Q (10^4–10^6) thin film resonators at 77K using high temperature superconducting films.

One of several basic designs is the wide microstrip resonator (Figure 9.8). It consists simply of a length of wide microstrip transmission line with the upper and lower sides made of a superconducting film deposited onto a sheet of insulating dielectric. In operation, standing waves are set up in this line section with a fundamental mode for which the length of the resonator is a half wavelength, determining the resonant frequency. Higher frequency modes can be driven with wavelengths of integer values of half-wavelengths. Microstrip resonators have high quality factors and so give high performance.

Another type of cavity is the coplanar resonator, again astonishingly simple in design and construction as can be seen from the illustration in Figure 9.9. It is made just by depositing a pattern of YBCO on to a substrate surface. It consists of two ground planes sandwiching a central strip, open circuit at each end, that forms the resonant section of a coplanar transmission line. The ground planes have silver evaporated onto each end to make contacts for the whole package. The two dimensional form of these resonators has advantages in relative ease of manufacture and integration into modern planar electronic circuitry.

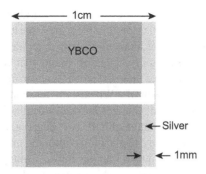

Figure 9.9   A high frequency coplanar resonator.

Superconducting Electronics in Space

Superconducting electronics play a crucial role in space communications for which a paramount requirement is low power consumption. In the depths of space, the background radiation temperature, the legacy of the "big bang" at creation, is about 3K: cooling may not pose a problem. Usually the power in a spacecraft is produced by solar cells, which constitute an important part of the payload. The cost of the launch of a spaceship is proportional to the total weight of the craft. Every watt of operational power saved as a result of using more efficient systems, including superconducting components, represents a smaller payload and substantial weight and cost saving.

In recent years, there has been an explosion in the use of radio, television, and mobile telephone communications. To enable continued growth, vastly greater signal capacity has been, and still is, needed. Along with the enormous increase in terrestrial communications, a huge expansion in communications through space has occurred. Large communications satellites (COMSATS) placed in the geosynchronous earth orbit have served large areas of the globe with television, telephones and other services. Video satellites (VSATS) and direct broadcast High Density Television (HDTV) are now commonplace. COMSATS use microwave transponder systems, which redirect signals from ground stations without reprocessing them. Rapid developments are taking place in space communications and geosynchronous orbit spacecraft are currently being supplemented by a new generation of intelligent digital switchboards, which handle traffic in a similar manner to a telephone switchboard. Superconducting electronics is one of the advanced technologies at radio and microwave frequencies being exploited in this communications revolution.

Superconducting Bolometers

A bolometer is a device for detecting radiation. The working part is a sensor, which should have a large temperature dependence so that it is capable of measuring a small temperature rise produced by incident electromagnetic radiation. The bolometer was originally invented by Samuel Pierpont Langley, an American scientist, astronomer and early enthusiast for heavier than air flying machines. Around 1880, while pursuing his interests in astronomy, Langley found that he could use very thin metallic foils, blackened to make them effective at absorbing radiation, as sensors. These were then placed in one arm of a Wheatstone resistance bridge, which was so sensitive that the extremely small changes in resistance that he was able to measure corresponded to a temperature change of only 10^{-5} degrees. Due to this high level of sensitivity, bolometers have many applications but they are especially important in astronomy. An early example is due to Langley himself who placed his equipment at the focus of a 13 inch telescope and was able to measure the variation of temperature across the Moon's disc.

Nowadays, high temperature superconducting bolometers are proving highly sensitive devices for measuring far infra-red and microwave radiation in the wavelength range of 20 to 100 microns. These bolometers make use of superconducting thin films deposited on an appropriate substrate. The resistance of the film can change dramatically when radiation is incident upon it, provided that the film is maintained at a temperature close to the mid-point of its normal-superconducting transition. This fulfils the criteria for bolometer design that they are able to give a large response to a small input signal, possess a low noise level and have a short response time. In addition, they can be made compact and easy to operate. Superconducting bolometers operating at the liquid nitrogen temperature are now replacing earlier photovoltaic cells, based on semiconductor technology, which become increasingly less sensitive and useful above 20 microns. Langley's visions for the use of bolometers in astronomy are now being amply fulfilled – indeed way beyond his wildest dreams. The far infra-red and microwave regions of the electromagnetic spectrum are currently of enormous interest to astronomers as evidenced by projects such as the COBE (Cosmic Background Explorer) satellite. The results obtained from such sophisticated research are transforming our understanding of the formation of galaxies in the early universe. Superconducting bolometers are playing an increasingly important role in these investigations.

Fabrication and Applications of Josephson Tunnel Junctions

The discovery of the Josephson effects and the fundamental ideas behind them has done much to deepen our understanding of superconductivity. Not only do the effects reveal entirely new and remarkable aspects of the physical world, described in Chapter 7, but also they have widespread applications in devices.

The earliest Josephson junctions, on which the first superconducting tunnelling experiments were carried out, were made in the form of metal–oxide–metal sandwiches. The general structure of this kind of device is shown schematically in Figure 9.10. To make it, a film of the superconducting metal (such as lead or tin) in the form of a strip of about 1mm wide, is evaporated onto a carefully cleaned substrate, which is usually glass. Each end of this filmstrip overlaps onto

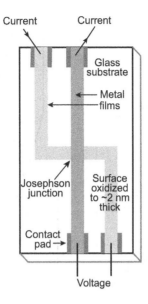

Figure 9.10 The structure of an early and much studied Josephson junction made by evaporating two strips of a metal such as lead through a mask onto a glass substrate. After the first strip (lighter grey) has been deposited it is lightly oxidized to coat its upper surface with a very thin (about 10^{-9}m) layer of insulator. The second strip (darker grey) is evaporated on top. The Josephson junction is the area formed at where the upper superconducting strip crosses the oxide layer on the lower strip.

pre-evaporated indium strip contacts. Next the film is very lightly oxidized to a depth of only a few atoms (about 2 nanometers) either by exposing it at room temperature to a controlled atmosphere of oxygen or by heating it in air. Then a cross strip of the same, or if required a different, superconductor is evaporated across the top of the oxide film. The device shown has four indium (In) terminals: the two at the top pass the current while the two lower ones tap the voltage developed. The room temperature resistance of such a device is of the order of 0.1 ohms. To make a tunnelling experiment, a current is passed through the junction and the voltage developed across it measured at about 1.5K – a temperature low enough for the lead (or tin) layers sandwiching the insulating oxide layer to be superconducting. The results are plotted to give the informative current–voltage characteristic of the type shown in Figure 6.11 and can be used to measure the superconducting energy gap.

Many types of junctions are in use. The technology for producing these "tunnel junctions" has become increasingly sophisticated as more and more applications have been invented and put into practice. A much used type of Josephson junction employs a weak contact between two superconductors. A "weak-link" is defined as a region across which it is possible to pass a resistanceless current that has a much lower critical value than those of the two superconductors it joins. The resistanceless current flowing through a weak link increases as the phase difference across the link increases; the critical current is that at which the phase difference reaches a maximum at 90° (as would be expected from the Josephson equation (1) given in Box 8 in

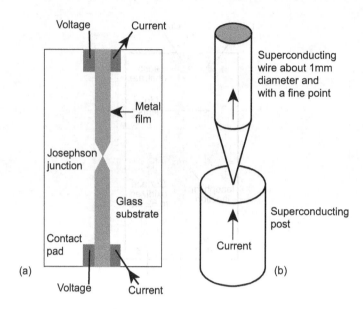

Figure 9.11 Two types of superconducting "weak links" which can act as Josephson junctions. (a) A very narrow neck etched out of a thin metal film. This limits the exchange of Cooper pairs. (b) A sharp point brought into contact with a flat surface; the area of the bridge between the two superconductors can be varied by the pressure at the contact, allowing the weak link behavior to be altered by simple adjustment.

Chapter 7). Weak-links can be made as a very narrow constriction (Figure 9.11(a)) or a point contact (Figure 9.11(b)) between two superconductors. They show tunnelling properties similar to those of the more standard Josephson junction (Figure 9.10). Weak-links made as constrictions or point contacts have the advantage of allowing easy access so that radiation can readily be coupled into or out of them. They are commonly used as practical radiation sources or detectors.

In high temperature superconducting ceramics, a weak link can exist as a region of poor contact between grains. Several advanced techniques have been developed to obtain them in a highly controlled manner for commercial use.

A remarkable feature of a Josephson junction is that it can sustain a current without a voltage being applied across it. *No energy is required to sustain that current.* But what happens if a voltage V is applied to a Josephson junction? Each superconducting electron pair will then gain an energy of 2 electron volts as it tunnels through the junction. The electrons in the pair now have energy they cannot use. This energy cannot be dissipated by scattering collisions in the junction. Under the influence of the applied voltage, the pair electrons have to rid themselves of the excess energy by radiating it away because they cannot accelerate by themselves as members of the condensed state. So the pair gives off a photon of energy E equal to the 2 electron volts gained from tunnelling. On account of wave-particle duality (Chapter 5) an electromagnetic wave (which has a frequency v given by E/h, where h is Planck's constant) is associated with the emitted photons and can be detected. Such a device which emits electromagnetic waves is actually behaving as an oscillator. So a Josephson junction acts as an oscillator with the frequency ν, when a voltage is applied to it. The frequency at which it operates depends linearly on the applied voltage (and is $2eV/h$). Putting some numbers in: if a voltage of one microvolt is applied, the frequency is 484MHz, which lies in a useful range not readily achieved using other devices. Maintenance of the constant voltage across the junction causes the amplitude of the supercurrent to oscillate in time: in addition to the usual dc current, there is an alternating-current effect. The process also works the other way round: radiation (in particular in the infrared or microwave regimes) can be absorbed by the junction exciting a voltage across it. In this way the device can be used as a detector of radiation. These "ac Josephson effects" were confirmed experimentally soon after their prediction by Josephson, and have since found a number of practical applications.

An elegant feature of the ac Josephson effect is that it allows voltage to be determined from a frequency measurement, which can be made with very high precision; hence it became for a while the basis of the U.S. legal standard volt. Another important application is that a measurement of the frequency as a function of voltage allows a value for the ratio e/h (electron charge divided by Planck's constant) of two fundamental constants to be obtained much more accurately (to about 1 part in 10^6) than could be achieved by any other method.

SQUIDs

Several years after the appearance of Josephson's papers, a lecture was given at Imperial College, London with the intriguing title: "SQUIDS and SLUGS – The Superconducting Zoo". These devices are practical consequences of flux quantization and phase coherence (see Chapter 7) around a superconducting ring intersected by a Josephson junction: these quantum effects are not just interesting scientific curiosities. Collectively they are the basis of operation of the SQUID, the acronym for "Superconducting QUantum Interference Device", now the most widely used small superconducting device. The great advantage of a SQUID is that it converts extremely small changes in magnetic field into a voltage measurable at room temperature. It is the most sensitive device of any kind available for the detection and measurement of minute changes of magnetic fields and in consequence is the foundation of an advanced measurement technology.

John Lambe and his colleagues at the Ford Scientific Laboratory first found quantum interference between two macroscopic superconducting waves and developed an r.f.-biased SQUID; largely because it was easy to fabricate it was widely adopted during the early stages of SQUID development. It consisted of a single Josephson junction in a superconducting loop (such a ring is illustrated later in this chapter (Figure 9.16)). This ring, containing a weak link, is coupled to a resonant circuit. The amplitude of the r.f. voltage across the circuit oscillates in response to the magnetic flux.

More commonly used nowadays is the dc SQUID, which consists of two Josephson junctions connected in parallel in a superconducting loop (Figure 9.12)

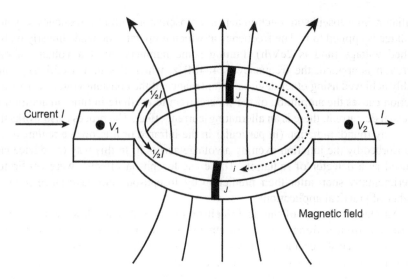

Current *I*

Magnetic field

Figure 9.12   The superconducting quantum interference device (SQUID) consists of a symmetrical ring with two arms each containing a Josephson junction *J*. A "measuring current" *I* is passed along the arms and divides equally so that ½*I* flows through each weak link *J*. When a magnetic field to be measured is inserted through the ring, it induces a second smaller supercurrent *i* which circulates round the ring adding to the measuring current through one weak link but subtracting from it through the other Josephson junction. A voltage *V*, which is measured, appears across the arms at the points V_1 and V_2. If the magnetic field changes, it causes the circulating current *i*, and in turn the voltage *V*, to alter and this is measured.

and operated by a steady dc current *I* inserted by an arm. This dc current splits into two equal parts (½*I*) each of which flows along one arm of the SQUID. These separate currents consist of Cooper pairs of electrons and tunnel through the junctions that act as weak links. This arrangement can be thought of as being analogous to Thomas Young's classic "two-slit" experiment that first showed optical interference between two beams of light. Interference established the wave nature of a light beam. In Young's experiment, a light beam is split in two by the slits and then the two beams recombined. If the two waves line up peak to peak, the beams reinforce each other and the resultant is bright, but if a peak lines up with a trough they cancel out and the result is darkness.

So Cooper pairs, like light, can show interference effects. Physicists have established that quantum interference is a ubiquitous effect of waves: it can be made to occur in beams of light, electrons, Cooper pairs, neutrons or atoms. Very recently, interference has been observed between two beams of "superfluid" liquid helium, a fluid that flows without viscosity and also behaves as a quantum wave. The result clinches the long standing belief, dating back to Gorter in the 1930s (Chapter 2), of analogous behavior between superfluid helium and superconductivity.

In a SQUID, the incoming Cooper pair wave current flows around the two halves of the ring, and recombines on the far side. The separate superconducting electron pair waves, travelling round each side of the ring, pass across the two Josephson junctions (which replace the slits in Young's optical experiment) and interfere when they meet. This quantum interference is a striking demonstration of the macroscopic phase coherence in superconductors introduced in Chapter 7. A magnetic field threading through the ring causes an additional persistent supercurrent *i* to flow round the ring (Figure 9.12); this combines in opposite senses with the supercurrents in the two sides of the ring, adding to the current on one side but subtracting from that in the other; the difference between the total current in each arm alters the synchronization of the recombining waves. When a magnetic field applied to the SQUID changes, the flow of the Cooper pair current in the ring alters. For a particular value of the magnetic field that threads the ring, the two waves line up peak-to-trough and the output is zero. A tiny increase in this magnetic field causes the two waves to line up peak-to-peak, producing maximum output; a further increase brings about the zero output again: the cycle repeats. The changes in current induced by variations in the magnetic field are measured as a voltage produced across the SQUID. Hence a SQUID is essentially a magnetic flux to voltage transformer of extreme sensitivity.

It is now nearly 40 years since the SQUID was first developed. John Clarke was one of the pioneers while working as a research student at Cambridge. At first the making of Josephson junctions was distinctly Heath Robinson. Early in 1965, John Clarke's supervisor, Brian Pippard, suggested that he might be able to make a SQUID and use it as a very sensitive voltmeter. During the traditional afternoon tea at the Cavendish Laboratory, a fellow student Paul Wright suggested that a blob of solder, which is a lead–tin alloy that becomes superconducting in liquid helium, deposited onto niobium might just conceivably make a Josephson junction. His idea was that the oxide layer, which ordinarily coats niobium, might act as a suitably thin tunnel barrier between the niobium and the solder. They hurried back to the laboratory, and in time honored fashion cadged a short length of niobium from a colleague, melted a blob of solder onto it, attached some leads and lowered it into liquid helium. To their delight, the device showed Josephson tunnelling. John Clarke indicates that it was important that Wright's idea worked first time: if it had not, they might never have bothered to try again. The device was named the SLUG, an acronym that parodied its appearance (a tidied up picture is shown in Figure 9.13). Later, Clarke went on to make a SQUID voltmeter that could measure 10^{-14} volts, a massive improvement of 10^5 times over the semiconductor voltmeters then in use.

In combination with a flux transformer, a SQUID can achieve femtotesla (10^{-15} tesla) resolution; this incredible sensitivity corresponds to about one part in 10^{10} of the earth's magnetic field. SQUIDs are by far the most sensitive detectors of magnetic fields known. The versatility of SQUIDs stems from their ability to measure any physical quantity that can be converted into a magnetic flux (that they transform to a voltage). This includes such properties as magnetic fields and field gradients, current, voltage and resistance. SQUIDs find applications in such diverse areas as medical diagnostics, geological prospecting, measuring instruments and in fundamental physics such as the search for gravity waves.

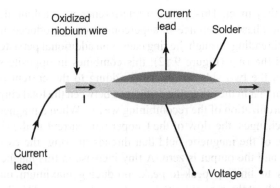

Figure 9.13   An early form of Josephson junction was the SLUG (superconducting low-inductance galvanometer) made from a lead–tin solder blob about 5mm long deposited on an oxidized niobium wire. The current I flowing through the niobium wire creates a circular magnetic field around it. When the two ends of the niobium wire are joined to form a super-conducting ring, the SLUG can be used as a sensitive magnetometer to measure magnetic fields. It has now been superseded by the much more sensitive SQUID.

High Temperature SQUIDs

The advent of the high-temperature superconductors has enabled SQUIDs to be made that operate in liquid nitrogen, a "warm" 77K! Exciting technological possibilities have been opened up by the availability of such devices. SQUIDs made from high temperature superconductors are simpler to use and have wider application than those built from the "conventional" superconductors. Some of the techniques used to produce Josephson junctions from conventional superconductors have been found to work for the new superconductors. It was quickly recognized that many of the early films obtained with the new materials had grain boundaries between single crystal regions in them, which could act as weak links and be used to make SQUIDs. Rapid advances have led to the production of high quality films with only a few grain boundaries. Sophisticated techniques have been developed to introduce artificial grain boundary junctions into the films in a controlled manner so as to produce reliable Josephson junctions.

Medical Applications of SQUIDs

One of the most important uses of SQUIDs is as a non-invasive probe in medicine. SQUIDs provide "windows" through which to peer into organs like the brain and the heart that produce minute magnetic fields. Using an array of SQUIDs, it is possible to map the spatial variation of the magnetic field produced by an organ, which can then be correlated with an abnormality. An important new technique, widely used in medical research, is "Magnetoencephalography" (derived from the Greek word

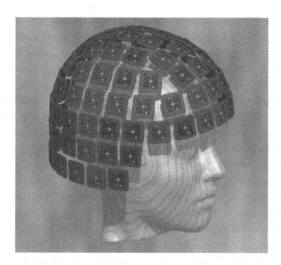

Figure 9.14 An arrangement of SQUID sensors for a 306-channel magnetoencephalo-graphy "helmet" (placed around a wooden head) that has been used in the Brain Research Unit at the Helsinki University of Technology. (After Hari and Lounasmaa (2000).)

encephalon for brain), generally abbreviated to MEG. It employs an array of SQUIDs to detect the tiny magnetic fields in the femtotesla (10^{-15}T) range generated from the weak electrical impulses transmitted between brain cells. The extremely small magnitude of these magnetic fields can be appreciated when it is realized that a femtotesla is some ten thousand million times smaller than the Earth's magnetic field. MEG is now increasingly employed by neuroscientists to study the way in which signals from our senses such as sight, hearing, touch, smell and pain are processed by the brain. For example, when we see a picture or hear a sound, signals are generated in the brain. MEG allows the simultaneous determination of the magnetic field profile produced by these processes and locates where the fields are generated in the brain. Mathematical modelling can then be carried out to assess what this magnetic field pattern corresponds to in terms of fundamental electrical activity in the brain resulting from the sensory stimulation.

The Brain Research Unit of the Low Temperature Laboratory at the Helsinki University of Technology is a world leader in the development of MEG for use in the medical field. It is led by Professor Olli Lounasmaa, a pioneer in the applications of SQUID technology in the 1960s. He has also pioneered techniques in ultra-low temperature physics making, among other discoveries, important contributions to the understanding of superfluidity in liquid helium 3. His life's odyssey illustrates the educational benefits of working in fundamental physics: Lounasmaa was quick to recognize the potential use of SQUIDs in fields outside his own rather esoteric discipline and he had the ability to exploit this – making important contributions in such disparate areas as brain research and superfluidity.

Modern helmet-shaped magnetometers are designed so that they can be fitted as close to the scalp as possible. A recent version of a MEG helmet made by Neuromag Ltd. for the Brain Research Unit at Helsinki is shown in Figure 9.14. This helmet has over 100 measuring points and uses more than 300 SQUIDS so that at each site, the magnetic field and the field gradient can be measured in every direction. This set up allows for an accurate determination of the magnetic field profile that in turn leads to a better understanding of how the brain works. By using such a helmet, the parts of the brain where particular senses are processed can be established. In addition, MEG studies show how these brain centers may operate. For example they have been used to locate the region in the cortex associated with hearing, and find out how it reacts to different sounds. As would be expected, work continues to go ahead to extend our understanding of how the brain responds to painful stimuli. The MEG technique is proving particularly useful for examining medical problems associated with the brain such as epilepsy and tumors. An array of SQUIDS close to the head of a patient suffering from epilepsy has been used to pick up tiny magnetic field fluctuations caused by the "electrical discharges", which characterize the condition. This makes it possible to pinpoint the location of the lesion in the brain responsible for the disorder, so that it can then be treated by neurosurgery.

In addition to establishing where particular sources of magnetic fields are located in the working brain, researchers have found that the strengths of these fields can also provide useful information about their sources. For example, studies are revealing how the sense of touch can be enhanced by training and practice. A research group working mainly in Germany, headed by Thomas Elbert, has made MEG studies of violinists and other stringed-instrument players. These musicians continually finger the strings with the left hand developing considerable manual dexterity and enhanced sensory stimulation. By contrast, much less dexterity is required by the fingers on the right hand operating the bow. MEG studies show that the magnetic fields produced in response to tactile stimulation of the fingers on the left hand are much stronger in such musicians than in non-musicians. Furthermore the response strength depends on the age at which the subject started to play an instrument: areas of the brain that process touch are capable of being reorganized through training.

The development of high temperature superconductors is having an impact on MEG technology. Since these new materials are superconducting above the normal boiling point of liquid nitrogen, SQUIDs fabricated from them require less insulation than the low temperature superconductors. This allows them to be placed closer to the brain.

Geophysical Applications of SQUIDs

High temperature SQUIDs can be used in much harsher environments than a medical laboratory; they are playing a vital role in geophysics research. They are now commercially available and are especially important in this application, since they frequently need to be used in remote areas. These devices operating above the boiling point of liquid nitrogen require little maintenance and are easy to transport. An array of SQUIDs provides the most accurate way of producing a "magnetic map" of a given region of the earth, which can help answer the question of whether there is any likelihood of oil and mineral deposits or sources of geothermal energy. SQUIDs are being used in measuring the surface impedance of the ground at different frequencies; this is known as "magnetotellurics". From knowledge of the surface impedance, it is possible to infer the underlying structure of the earth's crust. Such techniques are likely to become increasingly important as easily accessible resources continue to be used up. Archaeologists use a similar technique of impedance measurement, but on a greatly reduced scale, to map out areas of the countryside in the search for buried treasure and other remains.

Non-destructive Testing

A very different area where high temperature SQUIDs are coming into prominence is in non-destructive testing, especially in the aircraft industry. Flight safety is of paramount importance and aircraft are subjected to regular inspection and maintenance. Because they are exposed to frequent extreme changes of temperature and humidity, aircraft are prone to develop cracks and corrosion with potentially catastrophic results. For example, the fuselage of a plane is made from overlapping sheets of aluminium and the regions close to rivets in particular can easily corrode from deep inside. A non-invasive detection system is required. A mobile system for generating eddy currents is used in conjunction with a high temperature SQUID array. Analysis of the resulting magnetic field map enables the detection and evaluation of cracks and corrosion from regions well below the surface. This technique is currently being pioneered in the United States and Germany and is likely to have a large impact on routine aircraft inspection, especially as the reliability and robustness of high temperature SQUIDs improves.

Use of Josephson Junctions in Computer Technology

Soon after their discovery, it was appreciated that Josephson junctions could be switched very rapidly from the superconducting to the resistive state with very little power dissipation – creating the exciting possibility of developing a superconducting computer. Considerable time and money was spent in the research laboratories of IBM in making memory and logic circuits based on Josephson junctions. There were some early successes in producing logic circuits that could be switched in less than a nanosecond comparing very favorably with semiconducting devices. Furthermore Josephson junctions were put down on a chip; this attracted widespread acclaim and led to others entering the field, notably the Japanese. Almost all of the work carried out by IBM was based on Josephson junctions produced from lead alloys. Although these had good electrical characteristics, they suffered from being chemically unstable and changed with time. Despite having made considerable progress, IBM suddenly abandoned its superconducting program in 1983. One reason for this was the relentless progress being made in semiconducting technology making

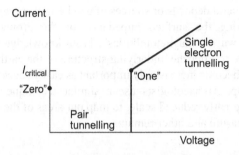

Figure 9.15 The Josephson junction as a bistable device. Once started to tunnel across a Josephson junction a pair current continues to flow with no applied voltage. If the current is less than the critical current $I_{critical}$, then the voltage across the junction is zero and the state is "**Zero**". If the current is increased above $I_{critical}$, the current flow becomes normal. There is now a voltage (typically a few millivolts) and the junction has switched to the state "**One**". To return to the "**Zero**" state, it is necessary to switch off the current. The jump "**One**" to "**Zero**" is fast: a few picoseconds.

superconducting computers less attractive. A further reason was their inability to produce a reliable memory bank for a main frame computer. Sadly, at about the time that IBM announced that they were to terminate their project, successful development of the niobium/aluminium-oxide/niobium junction was announced and a new "whole-wafer" approach devised. These overcame problems associated with the lead alloy devices and superseded them, having an important impact on subsequent developments and keeping the field active.

Using superconducting Josephson junctions in logic circuits has other advantages in addition to the low power dissipation and fast switching speed. These include the ability to incorporate superconducting connecting wires and superconducting grounding elements in logic systems. In addition, superconductivity is fairly insensitive to impurities and crystal imperfections so that it is easier to obtain materials with the required uniformity over a large area without having to grow large, high quality single crystals.

Despite attractive features, computers based on these devices have never taken off: the earlier ones needed to be run at liquid helium temperatures. Perhaps the introduction of high temperature superconductors will change that? As might be expected, it has given rise to an enormous impetus in development of a whole variety of logic elements. Many junction devices are made using thin films; they can be very small and so provide an attractive prospect for applications in large-scale integration circuits.

A range of uses, such as high speed switches, has stemmed from the fact that when the current exceeds the critical value, that is the maximum supercurrent I_c, a Josephson junction jumps to a state of finite voltage in which single particle tunnelling occurs. The current–voltage characteristic for a Josephson junction acting as a bistable device has two states "**Zero**" and "**One**", as shown in Figure 9.15. The state "**Zero**" corresponds to the situation that exists when a current flows with no applied voltage. To improve the sensitivity and minimize the switching time, this current is usually kept just below the critical current I_c of the junction. Any attempt

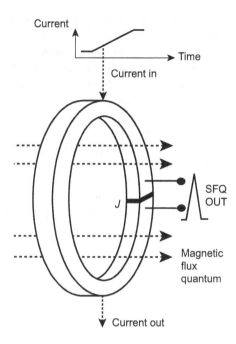

Figure 9.16 A superconducting loop interrupted by a single Josephson junction (J) acting as a rapid single flux quantum device can convert a current into a pulse. There are two stable states depending on the number of magnetic flux quanta threading the loop. When the slowly changing current (plotted at the top) reaches a threshold level, the system becomes unstable and flips in the short time of about a picosecond (10^{-12} seconds) to the second state by expelling a single quantum of magnetic flux from the loop. A short voltage pulse is generated by this flux jump.

to exceed this critical current I_c, causes the junction to switch reversibly into state "**One**"; this state is on the tunnelling characteristic for normal single electrons and there is a voltage across the junction, typically of a few millivolts. To return to the "**Zero**" state, the current is switched off. The jump from the "**Zero**" to "**One**" state and vice versa takes place in a few picoseconds (10^{-12} seconds). Switches as fast as this will find many applications in the future, especially in very high-speed computers. An advantage is that the energy required to switch is about four orders of magnitude less than that of the best semiconductor switches. The critical current depends on the size of the junction, the superconducting material and the temperature.

Amongst the many devices for integrated circuits and computers under continuing development are those based on RSFQ (Rapid Single Flux Quantum) logic. Konstantin Likharev pioneered this field at Moscow State University and has continued the work in the USA at the State University of New York–Stony Brook in a collaboration with IBM and AT&T Bell laboratories. The first RSFQ devices to be marketed are likely to be high speed analogue-digital converters. These logic elements are made from superconducting rings which contain a Josephson junction (Figure 9.16) shunted by a resistor. This active device can provide switching on picosecond (10^{-12}) time-scales and dissipates very little power. As we have already seen, the magnetic flux that can be contained in such a ring is quantized into discrete packets (Chapter 7). Hence one approach to building logic elements is to set up a superconducting loop enclosing a single flux quantum: then this represents the logic state "**One**"; absence of a flux quantum represents the state "**Zero**". In addition flux quantization can be used to store digital information, which can be inserted into or extracted from the loop extremely quickly. Every time a flux quantum enters or leaves the ring a very short voltage pulse proportional to the rate that the flux quantum enters or leaves the ring is produced. This "single flux quantum" pulse can be passed to other devices and used to make logic functions. RSFQ has been operated at a frequency of 360GHz, which is about a 1000 times faster than the Pentium computer. This fantastic speed far outperforms silicon technology with a very low power dissipation per gate operation – an important factor in the design of large, ultra-fast computers. Work in these and other Josephson devices proceeds at pace.

10 Organic Superconductors Enter the Race for the Top

Early Speculations: Could High Temperature Superconductivity Occur in Organic Compounds?

For many years, there has been speculation that organic materials might provide a possible route to high temperature superconductivity. Carbon is a most "talented" atom, which takes a leading role in many molecular productions including life itself. As a free atom, it has four electrons in its outer shell: just half the number needed to fill that shell. In forming compounds it shares these four electrons with other electron-sharing atoms, which can, and often do, include carbon itself, to make co-valent bonds. In combination with nitrogen, oxygen, hydrogen, phosphorus or other carbon atoms, carbon can be built up into an almost limitless variety of molecules comprised of chains, branched chains, sheets and composites of these. Some of these organic compounds are superconductors.

In the 1960s, William ("Bill") Little speculated that superconductivity at high temperature might be found in suitably designed organic compounds. He argued that an organic molecule might be devised that reproduces similar conditions for the formation of bound Cooper pairs of electrons to those found in metals. He suggested synthesis of a long molecule built of a chain of carbon atoms, connected by alter-nating single and double bonds so as to form a resonating "spine" along its length. One suitable molecule is illustrated in Figure 10.1(a). An electron put into this mol-ecule would be able to move freely along the length of this spine in a manner similar to that in a one-dimensional metal. To inject an electron, he proposed that on each side of the spine there should be molecular side chains extending outwards in a rib-like pattern. These side chains would contain a highly polarizable group of atoms such that electrons can move freely from sites at one end of each group to sites at the other end. This electron transfer (indicated by an arrow in the Figure 10.1(b)) creates a positive charge at the end closer to the spine. Little suggested that a suitable side chain molecule might be diethyl-cyanine iodide, available as a dye used to sensitize photographic emulsions.

Little's idea was that as an electron moves along the spine of his hypothetical organic molecule and passes each side chain, its electric field polarizes the side group and induces a positive charge at the end adjacent to the spine. As the electron moves

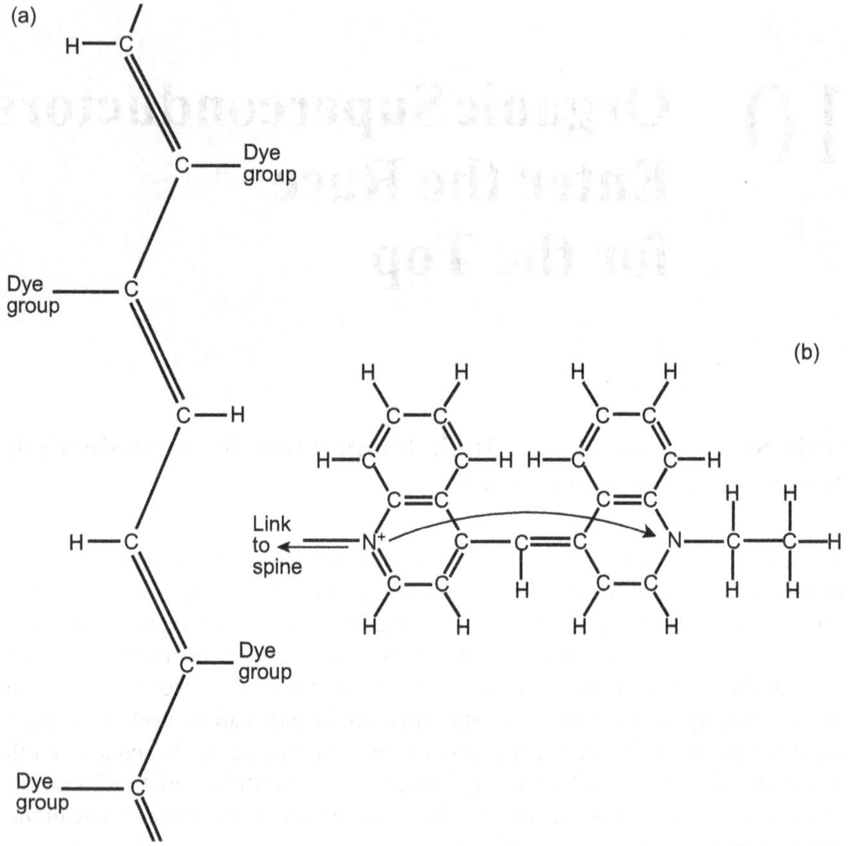

Figure 10.1 The hypothetical molecule that Bill Little suggested might be made into a super-conducting solid. (a) It is constructed from a "spine" of carbon atoms connected by alternating single and double bonds, thus forming a "resonating" structure. Electrons would be able to move freely along such resonating spine. (b) To inject carriers, side groups of the highly pola-rizable, common dye, diethyl-cyanine iodide, are joined periodically to the spine.

up the spine, it leaves a trail of induced positive charges in its wake thereby allowing a second electron to be attracted towards this region and hence to the first electron to form a pair. Detailed calculations carried out by Little suggested that this might occur at very high temperatures possibly at well over a 1000 degrees K. However, no organic compound made according to Little's prescription has shown any sign of conductivity, let alone superconductivity. His fascinating idea has yet to be realized.

Another Breakthrough: Superconducting Organic "Metals"

Until the 1950s, organic molecules had been regarded as insulating materials; that was, to change as organic conductors began to be found that showed metallic behavior – opening up a rich new domain of physics and chemistry. There was great excitement in 1973, when the organic compound TTF-TCNQ (tetrathiafulvalene-7,7,8,8-tetracyanoquinodimethane) showed anomalous resistance behavior around 54K, which was thought at first to be due to superconducting fluctuations. TTF-TCNQ is the best-known example of a group of so-called "charge-transfer" salts, which can be thought of as organic metals. The compound was first synthesized in the early 1970s by researchers at Johns Hopkins University; it consists of a donor molecule, tetrathiafulvalene (TTF), which is able to transfer electrons to an acceptor molecule tetracyanoquinodimethane (TCNQ) to form the charge transfer compound, TTF-TCNQ. This and related compounds are large molecules that are able to stack one on top of another, rather like pancakes, so that the transferred charge can move easily along the separate acceptor and donor stacks. Extensive research on the TTF-TCNQ molecule showed it to be the first organic compound found that demonstrated metallic behavior. The anomaly in the resistance around 54K was later found not to be due to superconducting fluctuations but instead due to a metal–insulator transition which could be attributed to the quasi-one dimensional nature of the material. There is no doubt that the work on TTF-TCNQ raised the profile and interest in this type of organic compound and led to considerable efforts to synthesize related compounds as vehicles for study of quasi-one dimensional systems.

Prominent among those carrying out research in this area was Klaus Bechgaard. Working initially at Johns Hopkins University, and then later at the University of Copenhagen, he succeeded in synthesising a new salt having the chemical designation $(TMTSF)_2PF_6$. Here, TMTSF the donor molecule tetramethyltetra-selenafulvane combines with the acceptor group PF_6 to form the organic compound. Further research showed that it was possible to produce a whole family of closely related organic compounds with the general formula $(TMTSF)_2X$ ($X = PF_6$, ClO_4, AsF_6 etc); these have subsequently become known as the Bechgaard salts and have quasi-one dimensional character. It was while Bechgaard was working on these materials in collaboration with the group headed by Denis Jerome from the University of Paris at Orsay in France, that genuine superconductivity was first observed in December of 1979 in an organic material at 1K. Investigations on $(TMTSF)_2PF_6$ revealed a transition from a conducting metallic state to an insulating form at around 12K, which could be suppressed by the application of hydrostatic pressure. This stabilized the metallic state that then went superconducting as the temperature was lowered to 1K. The discovery of a superconducting organic compound, some 15 years after Little's suggestion, has had an impact far beyond the finding of just one additional material in the list of those showing superconductivity. Investigations into other Bechgaard salts showed that it was possible to obtain superconductivity at ambient pressure at temperatures just above 1K. Describing these important events in a conference celebrating Organic Superconductivity Jacques Friedel could still recall after twenty years an excited Denis Jerome telling him on the telephone that he had at last observed superconductivity in an organic compound.

An interesting feature of the Bechgaard salts is their quasi-one dimensional character. There is growing recognition that they may be a case of a non-Fermi liquid of the Luttinger type (Chapter 11). This could be related to their being unconventional superconductors – especially since $(TMTSF)_2PF_6$ at low temperatures is in the

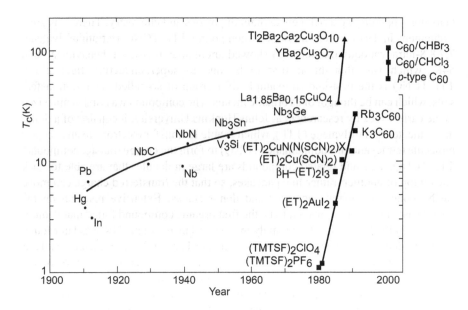

Figure 10.2 The rise of the critical transition temperature T_c over time beginning at the discovery of superconductivity by Heike Kamerlingh Onnes in 1911. A steep increase occurred in the 1980s following the breakthrough in the cuprates by Müller and Bednorz in 1986. The track for T_c in organic superconductors is shown to the right.

vicinity of a symmetry breaking, metal–insulator transition that would support the development of a one-dimensional non-Fermi liquid.

During the decade following the initial discovery of superconductivity in an organic material, there has been steady progress. An important advance was the synthesis of a class of organic metals with the formula $(BEDT-TTF)_2X$, where BEDT-TTF stands for *bis*(ethylenedithio)tetrathiafulvalene. This is generally abbreviated to $(ET)_2X$. The efforts made on organic materials were rewarded by a rapid increase in the critical temperature T_c during the 1980s (Figure 10.2). One of the $(ET)_2X$ materials, with X given by $Cu[N(CN)_2]Br$, has a superconducting transition temperature of 12K, which is the current record for this class of organic compound. Bechgaard and Jerome have remained in the forefront of this field; in 1991, they were jointly awarded the prestigious Hewlett Packard Prize of the Condensed Matter division of the European Physical Society and their published joint lecture admirably summarizes the status of the subject at that time. Despite intensive study, the pairing mechanism of superconductivity has not been conclusively established in organic compounds of the $(BEDT-TTF)_2X$ class. It is possible that electron–phonon coupling (see Chapter 6) could be relevant for organic superconductors. Unfortunately, the success achieved in the 1980s in raising the critical temperature T_c was not continued during the 1990s: the next major breakthrough is still eagerly awaited.

The relative lack of success in recent years in raising the transition temperature of organic superconductors has not diminished interest in the field of organic metals.

In fact, the level and quality of work in this field is highlighted by the award of the Nobel Prize for Chemistry for the year 2000 to Alan Heeger, Alan Macdiarmid and Hideki Shirakawa "*For the discovery and development of conductive polymers*". Their research has helped generate a widespread interest in the area of molecular and organic electronics, materials that are likely to be of increasing importance as this century progresses. Alan Heeger was a pioneer of the investigations into TTF-TCNQ mentioned earlier.

A Major Surprise: The Fullerenes – A New Form of Carbon

An exciting aspect of organic chemistry is its ability to generate compounds with completely new structures and unsuspected properties. Great progress has been made in the last decade in investigating a revolutionary class of organic materials known as the fullerenes. Since the nineteenth century it had been believed that the only pure forms of crystalline carbon are diamond and graphite. This situation changed dramatically in the second half of the 1980s with the unexpected discovery of molecular clusters of carbon. The background to this discovery is fascinating and illustrates well the interdisciplinary nature of much of modern research.

In the early 1970s, one of the research interests of Harry Kroto and his team at the Chemistry Department of the University of Sussex was in a certain type of long chain molecule called polyynes ($..C\equiv C-C\equiv C-C\equiv C-...$). Using microwave techniques it was possible to study their vibration–rotation dynamics. The first polyyne to be produced and investigated was cyanopolyyne HC_5N with a structure ($H-C\equiv C-C\equiv C-C\equiv N$). Elsewhere meanwhile a group led by Charles Townes had discovered in 1968 that black clouds smeared across the Milky Way contained long chain molecules. Townes had already become famous through being a joint recipient of the 1964 Nobel Prize for physics for work on quantum electronics which led to the development of the maser and laser. The importance of this later work by Townes and his colleagues is that it now appeared that space was no longer just the preserve of the astronomer but was also of enormous interest to the chemist attempting to look for and investigate exotic molecules. In collaboration with radio astronomers in Canada, Harry Kroto was able to detect cyanopolyyne in outer space. This discovery led to the subsequent synthesis and investigation in the laboratory firstly of HC_7N and then later HC_9N; both compounds were subsequently detected in space. The source of these long chain molecules in outer space was mysterious at the time, the most likely candidate being the aptly named red giant carbon stars. The success of these series of investigations led Kroto to ask whether it was possible to study carbon-star chemistry in the laboratory. It was in pursuit of this interest that led to his fruitful and far reaching collaboration with the group at Rice University in Houston, Texas headed by Robert Curl.

In the early 1980s, a small team at Rice University led by Richard Smalley had developed a laser-vaporization, supersonic cluster, beam technique that allowed small clusters of between 200 and 300 atoms of almost any element in the periodic table to be produced and studied. This was a major advance in cluster science and

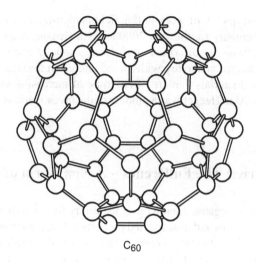

C_{60}

Figure 10.3   The Bucky ball: a new form of the element carbon that has the compositional formula C_{60} with a closed cage-like structure formed by linking 20 hexagons with 12 pentagons. This structure can be envisaged by putting a carbon atom on each corner of the sewn patches of a soccer ball, which is made from the same numbers of hexagons and pentagons.

was particularly important for refractory clusters since it made them accessible to detailed investigation for the first time. In this method, a pulsed laser is focused on to a solid target and vaporizes atoms of a refractory material enclosed within a gaseous environment, which is normally helium. Here the atoms re-aggregate to form clusters that are cooled by supersonic expansion, collimated into a beam, and detected using mass spectrometry. By using a graphite target and optimising the conditions, the group at Rice University, which had been joined by Kroto as a visiting scientist, was able to observe a high abundance of a particular cluster of 60 atoms, which is now known to be the C_{60} molecule; they also found a weaker signal corresponding to a less abundant C_{70} molecule. The question immediately arose as to the structure of such clusters. Discussions among the scientists, guided especially by Kroto's insight and knowledge, rapidly led to an elegant and intriguing answer: that C_{60} would probably be some sort of spheroidal cage.

Kroto recalled how immensely impressed he had been with the geodesic dome at EXPO 67 in Montreal. This dome had been designed by the architectural genius Buckminster (more commonly known as Bucky) Fuller who had pioneered curved geodesic structures formed by combining pentagons and hexagons. Bucky invented many things yet was always in dire poverty. His uncle, a high-powered businessman, became tired of Bucky always being broke, so he helped organize patents for geodesic, which were his nephew's latest invention, and arranged to exploit them. Bucky eventually became famous and a millionaire.

One of Bucky's domes became the model for the C_{60} molecule. Kroto suggested that C_{60} consists of a closed cage-like structure formed by linking 20 hexagons with 12 pentagons as shown in Figure 10.3. He called this new form of carbon, which can be envisaged by putting a carbon atom on each corner of the sewn patches of a soccer ball, buckminsterfullerene (if you need to say it aloud, see it as Buckminster-Fuller-ene). In popular jargon, the C_{60} molecule is frequently referred to as a Bucky ball. A whole family of closed caged carbon molecules, including C_{70}, has subsequently been found and has become known as the fullerenes. Comparisons of a C_{60} molecule with the modern soccerball proved to be particularly fruitful since the latter has isolated pentagons (usually black) separated by interlinked hexagons (white). It was known that structures with adjacent pentagons are unstable and this led Kroto to suggest that pentagon isolation one from another was the criterion necessary for stable carbon molecules. The smallest closed cage has to have 12 pentagons and the C_{60} molecule had to be the smallest structure with all the pentagons isolated (12 pentagons each with 5 sides). It had been shown already by the group that C_{70} could easily be formed from C_{60} by cutting it into two halves and inserting a ring of ten extra carbon atoms to produce a symmetrical egg shaped or rugby ball structure. Analysis of other fullerenes supported Kroto's viewpoint that pentagon isolation is the essential condition for stability.

Interest in fullerenes increased dramatically in the 1990s after Wolfgang Krätschmer and his team from Heidelberg in Germany showed that it was possible to obtain C_{60} and other fullerenes very easily from carbon soot by leaching them out using benzene; a graphite residue remains behind. They carried out this work in collaboration with a team led by Donald Huffman at the University of Arizona in Tucson. So Bucky balls, rare in space, can now be made relatively easily on earth in gram quantities. The resulting intensive worldwide investigation of C_{60} in particular, but other fullerenes as well, has given rise to some interesting science together with the possibility of technological applications.

Fullerenes Join the High T_c Race: Riding a New Elevator?

A surprising and significant discovery has been that certain salts of fullerenes are superconducting. The initial finding was made by Arthur Hebard and his group at the AT&T Bell Laboratories at Murray Hill, New Jersey. The nearly spherical C_{60} molecules form a close packed solid with a face centered cubic structure. In this solid there are three sites available into which it is possible to insert or intercalate other atoms or ions; three alkali or alkaline earth metal atoms A can go into these interstices in the structure to form a stoichiometric compound A_3C_{60} (Figure 10.4). The metallic state and superconductivity result from the transfer of electrons from the metal ions to the C_{60} molecule, which has a high affinity for electrons. Each alkali atom donates one electron to a C_{60} molecule: alkali metal-doped fullerenes of composition A_3C_{60} are ionic compounds, which conduct electricity. Hebard and his group showed that potassium doped C_{60} (which has the formula K_3C_{60}) exhibits metallic behavior and on cooling to low temperatures becomes superconducting with the rather high critical temperature T_c of 18K. Substitution of potassium by rubidium increases the superconducting transition temperature to 29K; T_c can be further increased to 42.5K by combining C_{60} with a rubidium–thallium mixture. These results have established intercalated C_{60} as the first three dimensional organic superconductor.

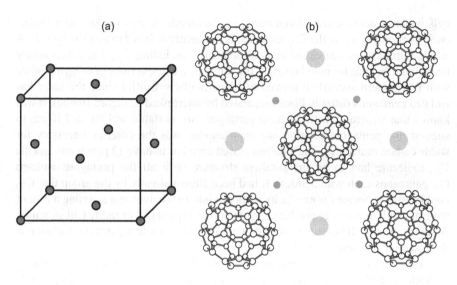

Figure 10.4   (a) The face centered cubic lattice. (b) The crystal structure of fullerene C_{60}: each fullerene molecule lies on a face centered cubic lattice point. In the case of crystalline A_3C_{60} compounds the alkali ions A occupy interstitially tetrahedral (large filled circles) and octahedral (small filled circles) sites between the fullerene molecules.

Superconductivity in doped fullerenes is currently an area of considerable research activity. A property of C_{60}, which favors the development of superconductivity, is that it has high-energy internal vibrations which can couple with electron states enabling the electron–phonon interaction needed to produce Cooper pairs. Although electron pairing does seem to be involved, the details of the mechanism responsible for superconductivity in the fullerenes are still by no means clear.

Next to a sorry tale. It has been suggested that superconductivity might well occur at higher temperatures in hole-doped than it does in electron-doped C_{60}. However, it is very difficult to remove electrons from C_{60}. (The concept of introducing holes by taking out electrons to make a conducting solid *p*-type has been described in Chapter 5.) One great boon of modern semiconductor technology is that it enables experts in device physics to use sophisticated tricks to achieve their desired experimental ends. Furthermore it is possible to engineer the critical properties of C_{60} in ways that are difficult to do with other materials. Late in 2000, Jan Hendrik Schön, Christian Kloc and Bertram Batlogg at Bell Laboratories reported that they had used a transistor device (an FET: field effect transistor) to inject significant densities of holes into crystalline C_{60}. Using this cunning technique, they claimed to observe superconductivity in *p*-type C_{60} with a critical temperature that depended on the number of holes and reached a maximum value of 52K. They made the exciting suggestion that T_c values significantly in excess of 100K might be achievable in a suitably expanded, hole-doped C_{60} lattice. By the middle of 2001, they stated that they had achieved that goal, reaching the critical temperature of 117K, by placing a crystal of C_{60} combined with tribromomethane $CHBr_3$ in the heart of a transistor. This achievement seemed to have resulted because the distance between the C_{60}

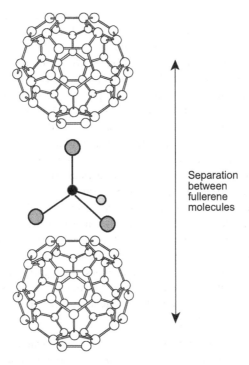

Figure 10.5 Increasing the spacing between the fullerene C_{60} molecules using tribromo-methane $CHBr_3$ molecules produces a crystalline solid with a separation of 14.45 Å. The larger distance between the fullerene molecules results in increase of the superconducting critical temperature T_c to an astonishing 117K. In the $CHBr_3$ molecule the black circle represents a carbon atom, the small light grey one a hydrogen atom and the three larger darker grey circles bromine atoms. However, this work is in doubt.

molecules was expanded to 14.45Å (Figure 10.5). Stretching out the spacing, increases the carrier density of states because more mobile charges are packed into a narrow energy range. Schön pointed out, "If the distance between the C_{60} molecules in the crystal is expanded, the carrier density of states is increased and T_c increases." This opens the door to exciting possibilities. Doped fullerenes could challenge the existing record for the highest T_c. If the distance between the C_{60} molecules could be increased to about 14.7 Å, then T_c should reach room temperature. Not only that: the technique of building C_{60} molecules into transistors could pave the way towards those elusive high-speed computers based on a super-conducting technology.

The work leading to the discovery of high temperature superconductivity in these *p*-type C_{60} materials was part of an ambitious program at Bell Laboratories studying organic materials. Schön, Kloc and Batlogg who spearheaded the investigations complemented each other well. Schön has a background in semiconductors and photo-voltaics while Kloc was a materials scientist with a talent for producing crystalline samples of complex materials. Both had worked at the University of Konstantz in Germany before coming to the Bell Laboratories. Batlogg, who was born in Austria and obtained both his first degree and doctorate at the ETH in Zurich, Switzerland, is a condensed matter physicist. Together with Robert Cava, he led a team that had carried out important work in the initial investigations of high temperature superconductors (see Chapter 3). Schön, Kloc and Batlogg and their colleagues produced a quite dazzling amount of research on organic materials resulting in a flood of publications in the most prestigious scientific journals. Many scientists around the world were amazed at the steady flow of seemingly very high quality research from the group. Most attributed it to the ability of the scientists concerned tied to the infrastructure of the Bell Laboratories, which has been in the forefront of scientific research, especially in condensed matter physics, for decades and can boast several Nobel Prize winners in physics. However, during the second quarter of 2002, things began to turn sour. Several people expressed concern about the validity of some of the data. It was noted that in several publications some graphs showed similar data despite the fact that the measurements had been made on different materials and varying geometries. Suspicions had been aroused that perhaps some of the data had inadvertently been duplicated in different publications, or it was rumored even fabricated. Commenting on the work of the group at Bell Laboratories, Laurence Eaves, one of the leading condensed matter physicists in the United Kingdom, told *Physics World* that there was general amazement at both the quality and diversity of the work adding,

> *"But no-one could understand how they could be so much better than the rest of us. Nobody seemed able to obtain results anything like theirs, but when you read their papers or heard their talks, it all seemed so plausible and exciting. I almost fell off my chair when I re-examined their papers and saw so many similarities."*

Bell Laboratories responded to such allegations by setting up an investigative committee headed by Malcolm Beasley of Stanford University, an eminent scientist who has worked for years in superconductivity. The high-powered panel also included Herbert Kroemer of the University of California at Santa Barbara who was a recipient of the 2000 Nobel Prize in Physics. They had a broad remit to establish the validity or otherwise of the whole work of the group.

For Jan Hendrik Schön the events of 2002 must have unfolded like a Greek tragedy. At the start of the year, he appeared to be the driving force behind a series of quite stunning discoveries with enormous scientific and technological implications. By the end of September the committee had found him guilty of forging data and he had been dismissed from Bell Telephone Laboratories for scientific misconduct, his glittering career in tatters. The findings of the committee set up by Bell Labs to look into allegations about his work concluded that he had faked experimental results in at least sixteen published papers. The extent of the deception is unique in condensed matter physics, a subject area not normally known for scientific fraud. The consequences of the fraud are considerable. Several people had tipped Schön to be a future Nobel Prize winner; he had already won the "Outstanding Young Investigator" award for 2002 from the Materials Research Society and was under consideration for a directorship of the Max Planck Institute for Solid State Research in Stuttgart. This ceased immediately following his dismissal from Bell Labs. The data fabrication and malpractice has been roundly condemned in many quarters. In *Physics World*, Art Ramirez, a physicist from the Los Alamos National Laboratory, states:

> *"Most physicists are dumbfounded. Schön broke many of the fundamental ethical rules of science, the basic one of which is "don't lie". His "misconduct" is an understatement. Many people have wasted time and money because Schön lied. Science is hard enough without being led down paths that others know to be fruitless".*

However, there are wider implications other than those, which affect Schön personally. All told he co-authored papers with some twenty people. What role did they play and should Schön alone be found guilty, which appears to be the case? Lidia Sohn from Princeton, who was among the first to draw attention to discrepancies in Schön's data, has addressed these points. Again in *Physics World* she states,

> *"It is amazing that 20 co-authors, one advisor and Schön's bosses at Bell Labs hardly questioned his work or even asked to see his devices working. That really mystifies me. To hear Schön say that he did not keep a lab book and that he "erased" his raw data is insulting to me as a scientist. Why did no one catch this from the beginning? I am totally dumbfounded."*

Chief in the firing line has been Bertram Batlogg, who was Schön's immediate boss. An anonymous contributor quoted in *Science* states: "Batlogg was certainly happy to bask in the glow when things looked wonderful. But you can't have it both ways and not accept some level of the burden of responsibility". The Beasley committee also asks questions over Batlogg's responsibilities when they stated, "Should Batlogg have insisted on an *exceptional* degree of validation of the data in anticipation that a senior scientist knows such *extraordinary* results would receive". In defending himself, Batlogg told *Physics World*,

> *"I have learned with the deepest of regrets that the verification measures that I and my colleagues have followed in this extraordinary case were not adequate to prevent or uncover scientific misconduct. As a result of this experience, I will apply additional and even more stringent checking procedures in the future. However, trust in colleagues shall and must remain one of the foundations on which we build future research and endeavors".*

There are lessons to be learnt by the scientific community. There is no doubt that scientists today are under enormous pressure to produce results. Much of the basis of science funding, and frequently of promotion and indeed employment itself, stems from producing a steady stream of scientific results and papers. A good *Curriculum Vitae* requires a large number of publications and there is a tendency for people to co-author papers for all sorts of reasons, which are unrelated to their actual contribution. Likewise the leading journals are all anxious to snap up the most important articles. The refereeing process tries to ensure that only *bona fide* work of good quality is published but the system cannot cope adequately with faked or plagiarized data unless this is blatant. In bygone eras, people carried out research to satisfy their own curiosity and to find out the workings of the universe. This is a far cry from the approach adopted by many scientists today where success is everything. It is probably not too surprising that in such an environment scientific fraud occurs.

In spite of this unhappy affair, the fullerenes are a new class of three-dimensional molecular solids that can be made to be both conducting and super-conducting and this has been a major advance. Optimism remains that related materials can be found that are superconducting at appreciably higher temperatures than yet found.

The discovery of the fullerenes and that of the first high temperature super-conducting cuprates took place within the space of a year of each other in the mid 1980s. It is likely that future historians will regard both these findings as among the most far-reaching scientific discoveries of the second half of the twentieth century. In Chapter 3, it was emphasized that the fact that Alex Müller was an IBM research fellow, enabling him to pursue almost any line of research that he wished, was a significant factor leading to the discovery of the high temperature superconductors. Likewise, Harry Kroto has frequently stressed the importance of allowing people to pursue "blue skies" research. This was succinctly put in the epilogue of a Friday evening discourse that he gave at the Royal Institution of Great Britain in 1996.

> "..... because research, today, is carried out under "applied" pressure and is imbued with the fear of failure. These factors are inimical to success in science for many scientists. The situation is exacerbated by the funding agencies, which monotonously drone on about so called "wealth creation" that few researchers have the luxury of really working in the dark. Only when you have no idea where the road leads does research embody the true spirit of scientific adventure and historically this approach has uncovered the true surprises and ultimately the great wealth-creating fundamental advances. ... ... Finally it is worth noting that C_{60} should have been detected 30–50 years ago by applied scientists studying sooty flames! Thus the discovery of the fullerenes is a shining testament to the power of pure fundamental science and serves as a timely warning that it can achieve results where applied science has manifestly failed."

The Nobel prize for chemistry in 1996 was awarded to Curl, Kroto and Smalley for their discovery of the fullerenes.

11 The Continuing Search for the Physical Origin of High Temperature Superconductivity

Ever since the discovery of the high temperature superconductors, one of the most intensively debated questions has been: why are their critical temperatures so much higher than those of the earlier superconductors? In principle, an answer might allow us to predict new superconductors. Understanding the origin of high temperature superconductivity presents a formidable challenge. Some years ago Sir Nevill Mott, questioning if there might be a generally acceptable explanation, concluded that, "There are probably as many theories as theorists". Today the situation has improved but a definitive theory is yet to emerge.

Since the BCS theory (Chapter 6) has had so much success in explaining conventional superconductivity, many have argued that it might at least make a starting point for developing a theoretical mechanism for superconductivity in the high temperature cuprates. A good example of this viewpoint can be found in the review written in 1989 by the eminent French theoretician Jacques Friedel. His article entitled "*The high-T_c superconductors: a conservative view*" argues that the classical weak coupling BCS theory is still valid provided account is taken of the additional complexity of sizeable antiferromagnetic interactions between electrons. This would imply binding the electron pairs by a phonon interaction but on the whole experimental evidence is not in accord with that.

Exotic Phases of Cuprates

A striking feature of the high temperature superconducting cuprates is that the parent or host materials are antiferromagnetic insulators: they have equal and opposite parallel spins on neighboring atoms (see Chapter 5). Electrical conductivity, necessary for superconductivity, can be induced in the parent materials by doping with an appropriate impurity element. As progressively higher concentrations of impurity atoms are added, a sequence of distinct phases with fascinating properties is revealed.

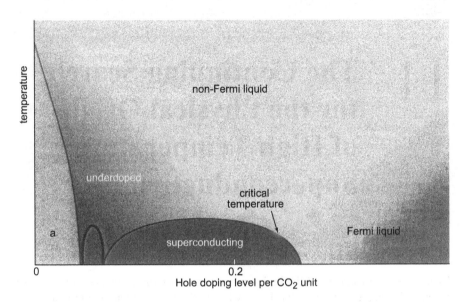

Figure 11.1 The way in which the properties of cuprates vary with temperature and doping (shown as the number of holes per CuO_2 unit). The phase labelled "a" on the left is **antiferromagnetic.** As the number of holes is increased, the antiferromagnetic phase disappears and a phase denoted as being **underdoped** appears. Further increase in the number of holes causes the material to act as a **non-Fermi liquid;** this is the phase that goes **superconducting** at lower temperature. The critical temperature is the line between the non-Fermi liquid and superconducting phases. If yet more holes are introduced, the material behaves as a normal **Fermi liquid.** In the right conditions, the small region between those of the antiferromagnetic and superconducting phases can comprise a **"spin glass"**. (After Batlogg and Varma (2000).)

As an example, we can consider the behavior with doping of lanthanum cuprate – the system in which Bednorz and Müller discovered high temperature superconductivity in 1986. Their original work was carried out on material in which some of the lanthanum was replaced by barium (see Chapter 3). Shortly afterwards it was found that strontium, another alkaline earth element, can be used instead of barium as the dopant; indeed strontium doped lanthanum cuprate can have a somewhat higher critical temperature. Like barium, strontium ions have a valency of two, compared with that of three for the lanthanum ions that they replace; such doping causes an increase in the number of hole-like carriers. It also plays a central role in stabilizing the different possible phases of the material.

The way in which the phases change with composition and temperature for the doped lanthanum cuprates is shown schematically in the phase diagram in Figure 11.1. Several different phases occur as the concentration of strontium is increased and temperature is varied. It is a reflection on the exotic nature of these materials that within different ranges of temperature and dopant concentration, the cuprates reveal insulating, antiferromagnetic, spin glass (and other complex magnetic orderings) and metallic phases, the last of which famously can be superconducting. In the doping range between the insulating and the superconducting regions, the magnetic behavior is complex. The undoped parent compound La_2CuO_4 is an antiferromagnetic insulator with a Néel temperature of 320K. However, doping by as little as 2 to 3% of strontium results in the formation of what is known as a spin glass.

It is worth turning aside for a moment to note that in themselves spin glasses have interesting and unusual physical properties. They occur quite frequently in a number of different systems and have been a source of interest since spin glass was first recognized as a novel magnetic state in the early 1970s. Typical examples of spin glasses include the noble metals copper, silver and gold doped with up to a few percent of a suitable impurity such as a $3d$ transition metallic element iron or manganese. Despite having randomly arranged magnetic moments, such materials undergo a transition to an ordered state below a characteristic temperature. This apparently paradoxical situation has been the focus of considerable experimental and theoretical research. For the cuprate $La_{2-x}Sr_xCuO_4$ the magnetic moments, which produce the spin glass, are on the copper ions that are also much involved in the development of superconductivity.

Addition of more strontium (7 to 8%) to $La_{2-x}Sr_xCuO_4$ produces a metallic phase, which persists until around 30% of strontium. The antiferromagnetism has now gone, and the material is a metallic conductor. At low temperatures this metallic phase is the one which is superconducting. The way in which the critical temperature alters as the number of holes is changed can be seen as the boundary line of the superconducting region in the schematic Figure 11.1. At an optimum doping of around 18% strontium, the superconducting transition temperature reaches its maximum value of about 40K. The behavior of the phase that goes superconducting is rather similar to that discovered by Bednorz and Müller for barium doped lanthanum cuprate ($La_{2-x}Ba_xCuO_4$) in which the superconducting transition temperature T_c rises as the amount of barium x is increased, until it reaches a maximum of around 30K for x about 0.15. Above 30% of strontium, the metal assumes a different character (behaving as a Fermi liquid) and superconductivity disappears. The material then becomes an insulator and superconductivity is suppressed. It is worth reminding ourselves that it was observations of the effect of doping in $La_{2-x}Ba_xCuO_{4-y}$ on the electrical resistivity and the development of a superconducting state at a higher temperature than previously found that led to a Nobel Prize for Bednorz and Müller and started the race for even higher transition temperatures in cuprate materials.

Bonding Between Atoms in the High T_c Cuprates

After its structure has been found (Chapter 4), the next step towards understanding how a material behaves is to establish how its atoms are held together. Before we can profitably discuss the bonding between atoms in the high temperature superconductors, we need to examine some aspects of the way in which electrons behave. In Bohr's early picture of the planetary atom (Chapter 5), it was assumed that each electron moved in some definite orbit as in the solar system. However, a fundamental consequence follows from the Heisenberg Uncertainty Principle: it is never possible to know exactly where a particle is; the best that can be done is to know the

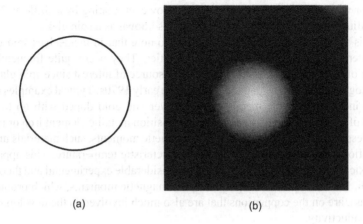

(a)   (b)

Figure 11.2   (a) Representation of an s-orbital as a sphere containing most of the probability of an electron in the s-state (quantum numbers ℓ and m both equal 0). (b) Artist's conception of the probability density $|\psi^2|$ of the ground state for the hydrogen atom. The probability density is spherically symmetric, greatest at the origin and decreases exponentially with distance r outwards from the origin. Each s-orbital can contain two electrons: one spin-up, the other spin-down.

probability that it is in any given place. The difficulties associated with Bohr's postulate that an electron follows a precise orbit were overcome in 1926 by solving the Schrödinger wave equation for the case of the hydrogen atom. A major triumph of the wave equation was that it showed directly the existence of discrete energy levels without any need to introduce arbitrarily the quantum conditions as Bohr had done in his theory of 1913. Furthermore, although Bohr's theory had given the correct values for the experimental optical spectrum for the hydrogen atom, it did not give correct energy values for other atoms and could not explain the way in which the brightness varied from line to line. The wave theory did give the spectral line energies as accurately as they could be measured experimentally and eventually was developed to describe the line intensities. In this sense the validity of the wave equation has been "proved".

In the wave theory, the position of the electron in the atom can be related to the probability of finding an electron, which is given by the square $|\psi^2|$ of the wavefunction ψ (a parameter which can be thought of as describing the motion of the electron wave). This probability that the electron can be found in a particular part of space varies from place to place. Although not strictly correct, a useful way of viewing ψ^2 pictorially is as a cloud – referred to as a charge-cloud. The density of this cloud at any point is proportional to ψ^2. Most of the negative charge is found at positions where the charge-cloud is densest (i.e. ψ^2 is largest).

Now the hopeless attempt to follow the electron on an orbit is abandoned; even the idea of a path has no meaning and is replaced by an electron charge cloud. The electron in an atom is now considered to reside within this cloud around the atomic nucleus. This charge cloud is called an *atomic orbital*, a reminder of its historical relationship to the previous idea that electrons follow orbits in atoms. The appearance

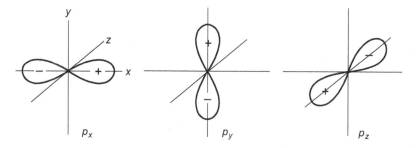

Figure 11.3 Representation of the shapes of the three p-orbitals for which the angular momentum quantum number ℓ is 1. The boundary lines contain about 90% of the electron probability density. Each orbital is distinct and they are mutually at right angles to each other. A crucial property of p-orbitals is their directionality. Each p-orbital can contain two electrons, so the three together can accommodate six electrons.

and size of an atomic orbital depends upon the actual energy state. For the lowest energy state of the hydrogen atom, the shape depends only on the distance between the electron and the nucleus – so that the charge cloud in this *ground state* is a sphere. Orbitals, which are spherically symmetrical like this, are called *s*-states; hence the electron is said to be in an *s*-state. For all atoms the lowest energy state is always an *s*-orbital denoted by the label (1*s*). Orbitals are generally drawn by showing a boundary surface, which contains most of the total electronic charge. So an *s*-orbital is shown as a sphere (Figure 11.2). According to the Pauli exclusion principle there can only be one electron in a state, and therefore an *s*-orbital can contain up to two electrons: one spin up, the other spin down.

To understand the behavior of high temperature superconductors, it is necessary to have a mental picture of atomic orbitals, and in particular, to recognize their symmetry properties. As we have seen, *s*-orbitals are all spherically symmetric. All the other types of orbitals are asymmetrical. The single electron in a hydrogen atom can be excited from an *s*-state into orbitals of higher energy with different shapes. A second type of orbital is that labelled *p*, which occurs when the quantum number ℓ is equal to 1; *p*-orbitals have the dumb-bell shape shown in Figure 11.3 and are discussed in more detail in Box 13. This type of orbital is relevant to the cuprates because bonding in the CuO_2 planes involves *p*-orbitals that exist on the oxygen atoms. Even more important are "*d*-orbitals" which exist on the copper atoms (when the quantum number ℓ is equal to 2). A description of these orbitals can also be found in Box 13.

In cuprates *d*-orbitals on the copper ions determine the bonding between the copper and oxygen atoms in the CuO_2 plane. The important fact that the cuprate superconductors all have a layered crystal structure composed of CuO_2 planes, often buckled, orientated perpendicular to the crystallographic *c* axis (Chapter 4) is due to the shape and directionality of the *d*-states on the copper atoms. The layers are separated from each other by planes made up of atoms of other oxides and rare earths. The bonding in the CuO_2 planes is of great significance because charge transport is mostly confined to these CuO_2 planes and it is the electrons in these planes that give rise to superconductivity. In the normal state, the electrical conductivity in the planes can be up to a thousand times greater than that in the direction perpendicular to them.

Box 13

Atomic *p*- and *d*-orbitals

In atoms, the orbitals are arranged in shells, labelled with the principal quantum number $n = 1, 2, 3$, and so on, in order of increasing energy (see also Chapter 5). In addition to n, there are other quantum numbers; these determine the number and the behavior of electron states in each shell for the hydrogen atom and those of the other elements. Additional quantum numbers are: the angular momentum quantum number ℓ, which can take integer values from 0 up to $n-1$; the magnetic quantum number m, which can take values $-\ell, -\ell + 1, \ldots, 0, \ldots, +\ell - 1, +\ell$. The first shell in an atom has a principal quantum number n of 1. In this case ℓ and m must both be zero. When ℓ is equal to zero, an electron is in an *s*-state. The fact that ℓ is zero means that an electron in an *s*-state has no angular momentum. The spherical shape of an *s*-orbital is indicated in Figure 11.2(a); the probability density is shown in Figure 11.2(b).

The second shell has a value of 2 for n. In this case, the quantum number ℓ can be 0 or 1. When ℓ is 0, m is also 0. When ℓ is 1, m can be -1, 0, or $+1$. Therefore the second shell *(n = 2)* of the hydrogen atom contains four orbitals: there are four states in this shell of the hydrogen atom, which happen to have the same energy (they are said to be degenerate): one of these states has an ℓ value of 0 and is a spherically symmetrical *2s*-state. The other three orbitals, for which ℓ is 1, are called *p*-orbitals. These three *2p*-orbitals have a boundary surface consisting of two regions which together resemble a "dumb-bell", as shown in Figure 11.3. One half of the dumb-bell is labelled with a positive sign and the other with a negative sign. It can be seen that these *2p*-orbitals, denoted $2p_x$, $2p_y$ and $2p_z$, are orientated mutually at right angles and have a very marked directional character. These three *p*-type orbitals are entirely equivalent, but are separate entities independent of each other. The Pauli exclusion principle allows each *2p*-orbital to hold two electrons: one spin up and the other spin down.

The next atomic shell is that labelled by n equal to 3. In this case ℓ can be 0, 1 or 2.

When ℓ is 0, there is one spherically symmetrical orbital called 3s. When ℓ is 1, there are three dumbbell shaped *3p*-orbitals with m equal to -1, 0, or $+1$. Orbitals for which ℓ is 2 are called *d*-orbitals; for these m can be $-2, -1, 0, +1$, or -2. So there are five *3d*-orbitals, which have the shapes shown in Figure 11.4. Like the *p*-orbitals, *d*-orbitals are strongly directional in space. For *d*-states the wave function changes from plus to minus and back again: the four lobes alternate in sign, two being positive and two negative.

The nature of the bonding between the atoms in the CuO_2 planes is indicated by the short distance of only about 1.9Å between each copper and its adjacent oxygen atoms. The copper and oxygen atom pairs are bound together by strong covalent bonds in which electrons are shared in pairs. A copper atom has four nearest neighbor oxygen atoms, each of which has three *2p*-orbitals (Figure 11.3) each providing an

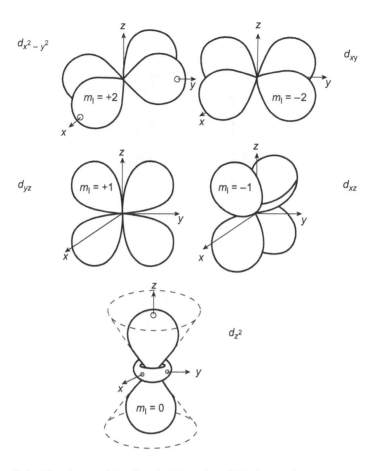

Figure 11.4 The shapes of the five d-orbitals for which the angular momentum quantum number ℓ is 2 so that the magnetic quantum number m can be $-2, -1, 0, +1$ or $+2$. There are five d-orbitals that can accommodate ten electrons. The relevant orbital for the copper atoms in the cuprates is ($d_{x^2-y^2}$) which bonds the copper to the oxygen atoms in the layer planes.

electron to a shared pair. It is generally agreed that the important orbital of each copper atom is that labelled as the $d_{x^2-y^2}$ state (that shown as the top left orbital in Figure 11.4). The $d_{x^2-y^2}$ orbital has four lobes whose axes lie along the x and y-axes in the CuO$_2$ plane. The orientations of the $d_{x^2-y^2}$ orbitals on the copper atoms and the $2p_x$ and $2p_y$ orbitals on the oxygen atoms in a CuO$_2$ plane are shown in Figure 11.5. These orientations allow the maximum overlap between the orbitals, which is in the x and y-directions in the plane; this overlap occurs along the line of the $2p_x$ (and $2p_y$) orbital on an oxygen atom and a $d_{x^2-y^2}$ orbital lobe of a copper atom. This maximum overlap produces the strong covalent binding (which is of the type known as a bonding) within the CuO$_2$ planes with each molecular orbital produced by the overlap containing its full complement of two shared electrons (Figure 11.5(b)). This means that there are strong single bonds between the copper and oxygen atoms in the layer planes as shown in Figure 4.1.

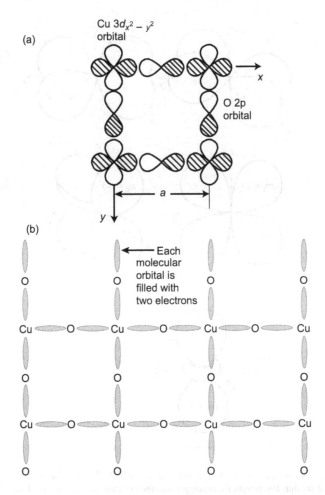

Figure 11.5 (*a*) The structure of the CuO$_2$ planes in the layerlike cuprates is comprised of almost square units with copper atoms on the corners and oxygen atoms centered on each side. The lobes of the $3(d_{x^2-y^2})$ orbitals of the copper atoms are directed towards the 2p-orbitals of the oxygen atoms. (*b*) These copper $3(d_{x^2-y^2})$ orbitals overlap with the 2p-orbitals of the oxygen atoms to produce hybridized molecular orbitals each of which contains a pair of electrons making a covalent bond between each copper and oxygen. This bonding produces the characteristic CuO$_2$ plane structure shown in Figure 4.1.

Electron Pairing in Cuprate Superconductors

An important breakthrough that took place soon after the discovery of the high T_c superconductors was the demonstration that superconductivity in the new ceramics, like that for conventional superconductors, also involved pairing of electrons. Pairing was shown by the experimental observation that the flux quantum Φ has a value of $h/2e$ (Chapter 7). Although this discovery that superconductivity in the new materials involves pairs of electrons was most significant, it posed the next important question: what is the nature of the paired state and what holds the pairs together? There is now widespread belief that the mechanism responsible for the formation of the Cooper pairs in some, if not all, high temperature superconductors is unconventional – in the sense that it does not involve an electron–phonon interaction (although not every-body accepts that).

In the case of conventional superconductors the BCS model (Chapter 6) provides the key to the resistanceless motion of electrons: they travel as Cooper pairs. The two electrons are bound weakly together with energy of the order of a meV; this attractive binding force between the electrons is produced by rapid interchange of virtual phonons, the process called phonon mediated attraction. The electrons are much more energetic and move much faster than the phonons. The BCS model depends upon the physical situation that the first electron has moved far away from the displaced ion when the second electron arrives and this reduces the negative repulsion between them. However, in the high T_c superconductors, the electrons and phonons move at more similar rates and the distance between electrons making up the pairs is less. The size of the Cooper pairs is much smaller and the coherence distance between the electrons is much less than that found for conventional super-conductors.

Philip Allen of the Department of Physics, State University of New York at Stony Brook has listed the arguments given in support of the view that an electron–phonon interaction is, at most, peripherally relevant to high temperature superconductivity. His comments carry weight and are considered in turn. He stresses that none are completely watertight.

(1) Most workers consider that the superconducting transition temperature T_c in the cuprates is too high for an electron–phonon mechanism. However there are others who feel that conventional theory might be less restrictive than generally thought.

(2) Up till now, there has been little evidence that the electron pairs in the cuprate superconductors could be held together by a phonon interaction. It is commonly taken as read that phonon coupling of electrons with d-wave symmetry into pairs is not favored. There is a counter argument: since the copper d-states are strongly directional and are essentially confined to the CuO_2 planes, d-wave pairing needs anisotropic coupling. Then in principle there could be strong phonon interactions for electrons. In addition, the cuprates have such small numbers of carriers compared with metals that conventional theory may need to be extended; this could lead to isolated bound pairs of electrons of a type referred to as bipolaronic, rather than the overlapping, long range (delocalized) pairs found in conventional super-conductors.

(3) It is widely held that for the cuprates in the normal state properties such as resistivity are not explained by electron–phonon interactions. But Allen categorically states that in its simplest form this is wrong: in fact, the temperature dependence of resistivity of the cuprates can be fitted well in terms of electron–phonon scattering theory in the usual way. If this is so,

then the arguments that are based on an unusual carrier behavior in the normal state may not be valid and cannot be used to exclude an electron–phonon mechanism for electron pairing.

(4) The fluctuating antiferromagnetic order, which is the source of exotic phases of cuprates discussed earlier in this chapter, strongly suggests that unconventional physics is involved. Although this does cast much doubt on electron–phonon mechanisms, the evidence remains largely circumstantial.

(5) The layered crystal structures of the cuprates (Chapter 4) indicates that special two dimensional physics is important. This is certainly true for electrons (and holes): the superconducting electrons are largely confined to the layer planes. However, the behavior of phonons in two dimensions may be little different from that in three.

In particular it is absence of direct experimental evidence for electron–phonon coupling in the high transition-temperature superconductors that has driven an intensive search for an alternative mechanism. However there are recent experimental indications of strong interactions between phonons and the electrical charge carriers. A. Lanzara and a group of his colleagues at Stanford University in California, together with workers in Japan, have examined the photoelectron spectra in three different types of copper oxide superconductor. Taking advantage of technical advances that enable fine details of the spectrum to be resolved, they have determined the energy dependence on momentum of the least energetic holes created by tuned photons from a synchrotron light source. These holes are the ones that are directly responsible for superconductivity. Distinct kinks in the carrier spectra have been observed. The kinks resemble those previously known in conventional metals that are interpreted as the signature of interactions between electrons and vibrations of the lattice, which result in an abrupt change of hole velocity and scattering rate near the phonon energy. Such a kink had been previously seen in one cuprate superconductor, but this new work suggests that it may be a general feature of these materials. If electron–phonon coupling does strongly influence the behavior of carriers in the high temperature superconductors, it should be included in any theory of superconductivity. But the many workers who reject the possibility that phonons provide the coupling for pairs are going to be hard to convince otherwise. They take the view that extraordinary claims demand extraordinary proof. After all, detailed theoretical calculations have given numerical values of the electron–phonon coupling strength that are large enough to be observed experimentally but are still much too small to account for the d-wave pairing in the cuprates. Nevertheless, the new studies cannot be dismissed easily, although the previous failure of electron spectroscopists to identify electron–phonon effects in the cuprates is hard to account for.

Probing High T_c Superconductivity Using Symmetry

There have been widespread efforts to find alternatives to an electron–phonon interaction as the cause of the formation of the Cooper pairs in the cuprates. One problem is that no experimental tests have been devised that can distinguish unambiguously between these various proposed sources of electron pairing. Examination of the symmetry associated with the electron pairs can go some way towards narrowing down the range of possible theoretical pairing mechanisms, which can involve a distinct symmetry.

Symmetry is one concept by which people throughout the ages have tried to understand the nature of the world and create order and beauty: a common theme of the sciences and the arts. Everybody is familiar with the idea of symmetry, but it can be a little difficult to decide precisely what is meant when something is said to be symmetrical. The idea of symmetry arose among the Greek philosophers and mathematicians as they sought to fathom the harmony of nature. In everyday language symmetry is an ally of beauty: it implies that an object is well-proportioned and well-balanced with a formal relationship between the parts and the whole. Beauty in women tends to be associated with symmetrical features! In Tudor times, Sir Thomas Browne surmised in his discussion of the symmetry of snow crystals, that nature possesses an order that we can aspire to comprehend. Scientific use is formal: a theory of physics is said to have a symmetry if its laws apply equally well after some operation, such as reflection, transforms parts of the physical system. By asking what operations will turn a pattern into itself, we can discover the laws that govern space and time. The natural symmetries of space and the physical laws define how atoms are bound together into a stacking pattern in a crystal, such as a cuprate, with an ordered translational arrangement in all directions.

Different states of matter have different symmetries. If the temperature of many physical systems is lowered, at some point they undergo a phase transition that typically results in a decrease or a "breaking" of some of their previous symmetries. A simple example can help us to understand the physics behind this reduction of symmetry, called *symmetry breaking,* at a phase transition. Consider water. The molecules are uniformly spread throughout a container, and the liquid looks the same regardless of the direction from which it is viewed: it is rotationally symmetric. Now watch the water as its temperature is lowered. Not much happens until the temperature reaches zero degrees Celsius; then a drastic change occurs. The liquid water begins to freeze, turning into solid ice. For the present purpose, the important thing to recognize is that ice is crystalline and, if any crystal is examined carefully, it looks different from different directions. The phase transition, at which the liquid changes to a solid, has resulted in a reduction in the amount of rotational symmetry: the symmetry is broken at the freezing point.

This is a common rule. When a phase transition occurs, such as that which takes place between a normal and superconducting metal on cooling below T_c, there is a symmetry change. In 1961 and 1962, while Brian Josephson was working as a research student at Trinity College, Cambridge, Phil Anderson gave a lecture course in which he introduced what was then a new idea of "broken symmetry" in superconductors. Josephson became fascinated by this concept and wondered whether it could be observed experimentally. He posed the simple question: "what is the physical significance of 'broken symmetry' in superconductors?" His own deep physical insight in considering this question led to his formulation of the Josephson effects.

Breaking symmetry drives changes of phase. The most familiar symmetry trans-
formations, such as the one for freezing water, are spatial: rotations or reflections.
However, arguments based on symmetry are by no means restricted to transforma-
tions of spatial objects. In 1954, Chen Ning Yang and Robert Mills found that impo-
sition of a symmetry at each point in space necessitated a new field, which is now
called a Yang–Mills field or gauge field. Gauge, which here just means measure, is
a term employed to describe a property of the field. An instructive analogy, which
does help in understanding gauge symmetry has been given by Heinz Pagels in his
book *The Cosmic Code*. Imagine an infinite sheet of uniform paper colored a uniform
shade of grey. The sheet of paper can be thought of as corresponding to a quantum
field. There is no way to tell where you are on the sheet of paper – it is said to be
globally invariant. This is true whatever the shade of grey: changing (i.e. "regaug-
ing") it would make no difference. Thus the shaded paper provides an analogy for
global gauge symmetry. A globally invariant field is undetectable, because it is the
same everywhere. Now imagine a sheet of paper that is colored in different shades
of grey. It is now possible to tell different regions of this sheet from one another, the
symmetry is broken: this sheet does not show global invariance. By analogy it is possible to determine the field effects when the symmetry is broken so that there are differences from place to place. A proper description of this rather abstract symmetry
really requires a mathematical treatment. A gauge symmetry expresses the fact that
the freedom exists to choose any path between the two end-points of the system without changing the energy difference between those end-points. This implies that we
can "regauge" a measurement of energy by any amount, that is we can shift the energy levels, leaving the differences unchanged. A field is said to show gauge symmetry if it stays unaltered following such a "gauge transformation".

At the superconducting transition it is the formation of the ordered quantum
mechanical state that breaks the symmetry because there is a change in the internal
symmetry of the type known as *gauge symmetry*. In the "conventional" superconductors at T_c there is a change in *global gauge symmetry;* a clear, but rather mathematical explanation, of this has been given by James Annett in 1995. In the normal
state of superconducting materials, global gauge symmetry exists above T_c but is
broken below when the ordered superconducting state is formed naturally. Thus
superconductivity is a state of spontaneous broken symmetry: the symmetry changes
naturally when a material is cooled below its critical temperature. High T_c superconductors are said to be "unconventional" because they show extra symmetry
changes in addition to this global symmetry transformation. Many workers have used
the extra symmetry, which is broken when an unconventional superconductor is
cooled below T_c, as a starting point for their search for an explanation of the origins
of superconductivity in the cuprates.

The extra symmetries in the normal state can include those of the crystal lattice
and spin rotation. A conventional superconductor is taken as one that retains these
fully below T_c, and in general, can be explained in terms of the BCS theory. An
unconventional superconductor is defined as one, which on cooling below T_c does
not retain completely the lattice or spin symmetries, which it had in the normal state.

Electrons in the conventional superconductors are assumed in the BCS theory
to form isotropic or "s-wave" pairs, having no net spin and zero angular momentum

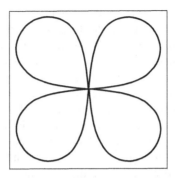

Figure 11.6   The shape of the electron pair wavefunctions of the anisotropic d-wave state, which may describe the electron pairs in high T_c materials.

(quantum number $\ell = 0$). These Cooper pairs have spherical s-wave symmetry (Figure 11.2); this means that the chance of finding one carrier in a Cooper pair with respect to the position of the other decreases at the same exponential rate in all directions in space: a sphere centered on one electron shows the probability of finding its partner. Conventional, or s-wave superconductivity, is the kind shown by most known superconductors. As we have seen in Chapter 6, one of the earliest triumphs of the BCS theory was the successful prediction that when a metal becomes superconducting, an energy gap opens at the Fermi surface. For an s-wave superconductor the BCS theory predicted that the value of this gap at zero temperature should be equal to $3.5kT_c$ (as discussed in Chapter 6). The simplest form of the model should be isotropic, that is, the same for all directions.

Some theoreticians working on the mechanism of superconductivity in the cuprates have favored mechanisms that are consistent with s-wave symmetry states. Many, but not all, of these theories tended to be modifications of the BCS phonon-mediated theory. From the beginning, it was apparent that BCS theory can accommodate other types of attractive interactions. For example, the s-wave symmetry of electron pairing in conventional superconductors might be replaced by more complicated symmetries such as p or d wave. However the majority view is that it is quite unlikely that the BCS mechanism could produce superconductivity at the high temperatures reached. A new mechanism is needed. So other theorists have emphasized mechanisms that generate so-called d-wave symmetry states. In the last few years, strong experimental evidence has been presented that the superconductivity in the cuprates is indeed unconventional, having d-wave symmetry in which the electrons pair with non-zero orbital angular momentum. For a start, the layer structure of the cuprate superconductors suggests that the Cooper pairs in these materials may well have anisotropic d-wave symmetry (Figure 11.6) with an angular momentum quantum number $\ell = 2$. In particular much attention, but by no means all, has been focused on the $d_{x^2-y^2}$ pairing state; in fact, d-wave symmetry is implied by a number of suggested pairing mechanisms, especially those involving magnetic interactions, which we have seen are known to be of importance in the cuprates. The electron pairing in a d-wave superconductor is highly anisotropic, the $d_{x^2-y^2}$ pairing state orbitals look like a four-leaf clover with alternate positive and negative lobes (Figure 11.6). Each lobe represents a likely position of one member of the Cooper pair with respect to its partner. However, a few experiments have consistently disagreed with this picture favoring instead the isotropic s-wave pairing that describes conventional superconductors. One of the possible reasons for the confusing and sometimes contradictory experimental results is that the cuprates are so difficult to work with. Conclusive experiments need high quality crystals with known doping levels achieved by control over their oxygen content. However, only small crystals with poor mechanical strength are available at the moment. Measurements for cuprates are very difficult to disentangle from various possible defect and impurity effects. Nevertheless, symmetry tests can help to narrow down the possible pairing mechanisms. Establishing if d-wave symmetry is characteristic of the cuprates would place strong constraints on theory. One problem is that the $d_{x^2-y^2}$ and s-waves can readily mix: a model which takes into account both the coupling between the planes and chains, blending both s- and d-wave features may provide a natural explanation for all the contradictory experiments.

The technique of high resolution photoemission spectroscopy has been used to measure electronic structure in the superconducting state of Bi2223 (T_c = 110K) and Bi2212 (T_c = 85K). The reduced energy gaps have been measured as about $6.5kT_c$ and $6.1kT_c$ respectively, about twice that predicted by BCS for conventional s-wave superconductors. The experimental results show that the energy gap is strongly anisotropic, a feature consistent with d-wave pairing. Although a microscopic theory of high-temperature superconductivity may still be some time away, agreement on the symmetry of the electron pairing state would be an important step forward. A d-wave nature of the pairing is supported by many experiments (although these are not easy to perform and difficult to interpret) and has been widely accepted. A readable account of experiments, which have added to the already convincing evidence that YBCO has an unconventional pairing state with $d_{x^2-y^2}$-wave symmetry, has been given by John Kirtley and Chang Tsuei in *Scientific American*.

So it seems likely that superconductivity in the cuprates is associated with d-wave symmetry. In that case, BCS theory based on s-wave symmetry cannot provide a viable explanation. The question of symmetry of superconductors remains an important area of research, particularly since superconductors with strongly anisotropic order parameters also show such a range of interesting physical phenomena.

A Path Towards Understanding the Mechanism of High T_c Superconductivity

In the excited scramble that followed the discovery of high T_c superconductivity in the cuprates, many had a vision of their particular expertise providing the lead to an understanding of the working mechanism. Many possible pairing intermediates were suggested, among others are spin waves, polarons or excitons. How to reduce the possibilities? The first task was to remove the ones that did not conform to valid experimental observations or reasonable theoretical ideas. Following its recognition in about 1993, d-wave symmetry seemed to provide one way; however, it was not to prove as useful as first thought because many of the proposed pairing mechanisms can be actually made consistent with d-wave symmetry.

Sir Nevill Mott and P.W. (Phil) Anderson, two of the three recipients of the 1977 Nobel Prize for Physics, had conflicting viewpoints on the mechanism of high temperature superconductors which have been interestingly summarized in two letters to *Physics World* in January 1996.

At one time, the bipolaron model put forward by the late Sir Nevill Mott and his colleagues had some prominence. A polaron describes the lattice distortion, which accompanies the motion of an electron through a solid. Bipolarons are bound pairs of polarons, which are mutually attracted by the lattice distortion. Using this concept, it was possible to explain many experimental observations, notable among them being the small values of transport properties along the c-direction that is perpendicular to the CuO_2 planes and also the short range of the pairs. However, the fact that the cuprates are conductors that have a Fermi surface militates against this kind of model.

Phil Anderson's approach is most radical. He argues that the high temperature superconductors represent a new type of condensed matter whose understanding requires fundamentally new physics. To carry this forward is a massive undertaking. As a set of guidelines which logical argument should follow towards an acceptable theory, he has laid down a set of *"central dogmas"*, which may be thought of as scientific analogues of a declaration of human rights! These *dogmas,* based on well established experimental data, map out a path for theoreticians to follow.

Anderson's *first dogma* refers to the knowledge that the carriers of both electricity and spin reside in the CuO_2 planes. These carriers arise in the hybridized $O(2p)$–$Cu(d_{x^2-y^2})$ states, which play the major role in binding together adjacent O and Cu atoms in the CuO_2 planes (Figure 11.5). Thus, a path taken towards a theory should be restricted to considerations of those physical effects, which originate from the electrons being confined to these planes.

The *second dogma* is that high T_c superconductivity and magnetism are closely linked in the cuprates. An important consequence of the first two dogmas is that they restrict the theory to a single electron band of $(d_{x^2-y^2})$ symmetry.

The purpose of the *third dogma* is that the theoretician should keep his attention focused on only one interaction of these electrons, which is taken to be the one large enough to open up an energy gap and lead to superconductivity. This is analogous to the BCS approach of assuming that only the pairing interaction which leads to superconductivity need be considered.

Attention to these first three dogmas severely limits the number of possible theories. Anderson warns that there is a danger of theoreticians trying to describe all known superconductors in terms of the BCS theory and its extensions. The phonon-based pairing mechanism developed by Bardeen, Cooper and Schrieffer (Chapter 6) for low-temperature superconductors requires the Coulomb repulsion to be described in terms of so-called Landau–Fermi liquid theory, the basis of the model for metals outlined in Chapter 5. A metal is usually treated as an electron gas. In a Landau–Fermi liquid, the electron and hole excitations of a metal carry both the charge and the spin. The charge carriers are not simply the electrons but are modified (or "renormalized") by interactions with other electrons to form what are called "quasiparticles". The properties of the material can then be understood in terms of weak interactions between these quasiparticles.

Anderson has pointed out that many superconductors do not fit readily into this scheme but should be thought of as deriving from metal-like solids which do not behave as a Landau–Fermi liquid. In addition to the cuprate and organic super-conductors (Chapter 10), examples of such materials include Nb_3Sn and $PbMo_6S_8$ (one of a class of superconductors which are ternary molybdenum chalcogenides some of which have potential applications because of their very high critical fields). Also coming into this category are the intriguing heavy fermion superconductors, such as UBe_{13}, $CeCu_2Si_2$ and UPt_3, which are characterized by having what is known as a very large "effective mass", about 100 or more times that of the free electron mass so that the carriers have a high inertia. There is a distinct possibility that all of these types of materials have unconventional symmetry and an unusual pairing mechanism. An important point is that breakdown of the Fermi liquid model can occur near a symmetry-breaking transition, such as the transition from paramagne-tism to antiferromagnetism which occurs in the heavy fermion metals. This radical direction put forward by Anderson is being followed up by many groups but others remain sceptical.

A theory of superconductivity must rest on understanding the metallic state from which it develops. In general, the Landau–Fermi liquid theory gives a good explanation of the properties of metals and the BCS theory of conventional super-conductivity which relies on it as a starting point. However, the cuprates may not fall into this category. They may depart from usual metallic behavior in the normal state. They also seemed to have posed the puzzle of an absence of electron scattering by phonons in the normal state. However the experimental situation is by no means clear because the way in which their electrical resistance change with temperature does seem to fit in with phonon scattering, as Philip Allen has noted. Anderson takes the view that the carriers in cuprates in the normal state above the superconducting transition do not behave at all like a traditional Fermi liquid. What enables this breakdown of the Landau–Fermi liquid theory? It can take place near a symmetry breaking transition, such as that from a paramagnetic to an antiferromagnetic state which the cuprates can readily undergo.

The next question is: what is the nature of the resulting electron liquid in the cuprates? Anderson has revived interest in the so-called "Luttinger liquid" which was originally put forward in the 1960s to explain certain properties of one dimensional conductors, but which he contends may also be applicable in two dimensions, typical of cuprates. Anderson believes that the experimental observations strongly support a *fourth dogma:* that the normal metal state in cuprates is a two-dimensional Luttinger liquid. Experiments show the absence of any coherent electron transport perpendi-cular to the CuO_2 planes, i.e., in the c-direction above the superconducting transition temperature in the normal state. This would be expected from Anderson's approach but is inconsistent with the properties of a Landau–Fermi liquid. A curious property of the Luttinger liquid is that the spins and charges are carried separately by particles called "spinons" and "holons" respectively. This results in the development of a remarkable metallic state in which the electron dissolves into its magnetic and conducting components. The *fifth dogma* is that the two-dimensional state has separation of charge and spin into excitations whose effects are only meaningful in the plane. Finally, *the sixth dogma*: interlayer hopping makes an important contribution to the superconducting energy.

The introduction of these six dogmas has provided an outline for the direction that a viable theory could take. At present, much work is in progress on each part of the route. Yet in spite of the enormous worldwide effort by theoreticians, we remain in the frustrating position that there is still no consensus about how high T_c superconductors work. The solution to the problem of the cuprates may well lie in the brave new world of non-Fermi liquids. A major spin-off of the work on high T_c superconductors is that the theory of non-Fermi liquids has become an important growth area in condensed matter physics, and is likely to have important applications in understanding many new and exotic materials showing properties that do not fit easily into well-established theoretical ideas. Certainly, the ongoing studies in this area prompted by the urge to understand superconductivity in cuprates are proving most stimulating and productive of original ideas, and are themselves revolutionizing our understanding of solid state physics.

Unity or Diversity: More Than One Cause of Superconductivity?

Superconductivity stands at the edge of the known world. Researchers in the field are modern day explorers of the blank spaces beyond. The area containing the conventional superconductors is now well mapped by BCS. However, an explanation for superconductivity in the region comprising the unconventional high-temperature cuprates remains elusive, although the charts held by the workers on non-Fermi liquids seem to point to a second and new trail to follow. But are we travelling paths giving distant views of what in fact is one reality? Can there really be different sources of the quantum world of superconductivity? Or is there a single unifying interaction that can be plotted in a single theory encompassing all materials in which it is found?

Coupling between electrons and phonons (lattice vibrations) binds the electron pairs that are responsible for conventional superconductivity. The very recent photoelectron spectroscopy experiments may indicate a surprising similarity of the unconventional superconductors to their conventional low-temperature counterparts. Certainly such observations have reinvigorated travellers towards unification. At the moment, it is established that electrons in unconventional superconductors are bound in pairs, but the binding force remains unknown. The interaction responsible is still a matter of debate: neither experimentalists nor theorists have been able to provide a convincing picture of what is happening. How exciting to tread unknown terrain having what the Abstract Expressionist painter Willem de Kooning called "a slipping glimpse" of that new world!

Epilogue:
the Way Forward?

The old Chinese proverb exhorts "May you live in interesting times!" The decade following the initial discovery of high temperature superconductivity has been a fascinating and exciting time for many scientists and engineers exploring this mind stretching new area. Research on the new superconductors has been a truly global effort. Much of the fundamental and applied work has been carried out within Europe, America or Japan but there have also been significant contributions from the countries of the former Soviet Union as well as from China, India, Australia, New Zealand, South Africa, South America and elsewhere. Equally striking is the wide range of expertise of the people who have made a significant input. For many years, superconductivity has been regarded as almost the sole preserve of the physicist and electrical engineer. By contrast, scientists and engineers from a wide range of disciplines have contributed towards our understanding of high temperature superconductivity.

Shortly after the discovery of the new superconductors, an editorial appeared in the journal *Nature* with the intriguing title *"Chemists come in from the cold"*. Its author, John Maddox argued that the studies required to understand high temperature superconductors have given an enormous boost to theoretical chemistry. He suggested that it is chemists who understand most about parameters such as atomic radii, bond lengths, valence, co-ordination numbers and electronegativity, which are very much in the domain of the chemist, but are central to understanding the new superconductors. Among chemists working in the field was Linus Pauling, a most distinguished scientist and winner of the 1954 Nobel Prize for chemistry and also that for peace in 1962. He developed a theory for high temperature superconductivity, which relied heavily on his formidable chemical intuition.

The leading countries in research and development in high temperature super-conductivity are undoubtedly America and Japan. Commentators have implied that a race exists between these two nations for technological supremacy. In the last thirty years or so Japan has dominated the world market with consumer products such as cars, televisions, cameras, video and digital cameras, the walkman, recording equip-ment, and so on. The fear has been expressed in some quarters that application of high temperature superconductivity is the last chance for America to prove its tech-nological strength and that there is every likelihood that yet again it will lose out to Japan who could corner the world market. Shoji Tanaka, one of the leading figures in the superconductivity research effort in Japan, has described the frenetic activity in his country during the first year of research into high temperature super-conductors. This effort, which involves a heavy investment in both manpower and funding, continues to this day, and Japanese scientists have made substantial progress in producing good quality superconducting wires and thin films suitable for commercial exploitation. Insight into their progress can be obtained from a series of articles published in the journal *Superconductor Science and Technology* (Vol. 13, January 2000). The wide variety of articles presented, as well as their technological sophistication, emphasizes the bright commercial future of high temperature super-conductivity during the new millennium with Japan much in the forefront.

Nevertheless the distinguished physicist John Rowell paints a reasonably optimistic picture for America's success in the exploitation of high temperature superconductivity. Rowell has spent virtually his entire career investigating super-conductivity; he carried out seminal work at the Bell Telephone Laboratory supporting the ideas of Josephson, as described in Chapter 7. Until recently, he was at Conductus Inc. of Sunnyvale, California, one of several relatively small high tech-nology companies set up to research and develop the commercial exploitation of high temperature superconductors primarily in the area of superconducting electronics. Few are better placed than Rowell to comment on the overall commercial prospects for applied superconductivity, an area on which he has lectured and written about quite extensively. He has argued that many of the important discoveries originated from America where there has also been rapid follow up. In 1988, Rowell analyzed the published papers from five important conferences featuring superconductivity, one just before the discovery of high temperature superconductors and the remainder in the period shortly afterwards, and concluded that during the initial years America was certainly holding its own against Japan in this field. Rowell has pointed out that when type II superconductors were being developed in the 1960s, many large American companies were actively involved. Nowadays, there are far fewer, pos-sibly only IBM, AT and T and DuPont. Much of the current commercial exploitation of high temperature superconductors is taking place in small companies. Rowell has expressed fears that adequate funding might not be available for product develop-ment. He believes that enormous progress has been made in the fundamental investi-gations but much more work is needed to get commercially viable products into the market place. Failure to capitalize on early successes could well result in America again losing out commercially to Japan in what has been predicted to become an annual 20 billion-dollar industry during the early years of this century.

John Crow, at the time Director of Science Policy Research at Iowa State University, has reached a similar conclusion. He argued that Japan was quickly able to develop a national strategy to capitalize on high temperature superconductivity whereas the effort from America was much more fragmented. The greater emphasis on individuality, which characterizes America, might be a stumbling block to a project such as the commercial exploitation of high temperature superconductors, which lends itself to a focused and co-ordinated national effort. Such a directed national effort is better provided for in Japan through the Ministry of International Trade and Industry of the Japanese government.

John Rowell has been part of a team developing a "*Technology Roadmap for the Superconducting Electronics Industry*". The purpose of a Technology Roadmap is to define the medium and long-range needs of industry; in recent years, such plans have become part of the American commercial scene. In an earlier case study for the semiconductor industry, emphasis was placed on the empirical "Moore's Law" and this is also likely to be useful for the superconducting industry. In 1965, Gordon Moore of Intel predicted that the capacity of a computer chip would double every year. This was based on an analysis of the price/performance ratio of computer chips over the previous three years. Although it is a purely phenomenological observation, it has proved to be remarkably reliable and is now embodied in "Moore's Law" which predicts that capacity will double roughly every eighteen months.

Several aspects of superconductivity are under consideration for producing a "Technology Roadmap". These include materials, thin films, wireless applications, NMR and MRI, SQUIDS, digital applications, and refrigeration and cryopackaging. Rowell believes that by the year 2006, twenty years after the initial discovery of high temperature superconductivity by Bednorz and Müller, applications sited in these areas should be up and running in the market place, supporting a viable super-conducting electronics industry. If this is not the case and by the year 2006 super-conducting electronics is still at the Research and Development phase, Rowell believes that interest will wane and insufficient funding will be made available for continued work. He draws attention to the likelihood for commercial opportunities for hybrid superconducting/semiconductor electronics. Rowell points out that industry in general has an aversion to cryogenic temperatures. At the moment, a great effort is in progress to make simple, efficient and reliable refrigerators capable of operating around 60K for the high temperature superconductors or about 10K for the type II superconductors. For some while it has been realized that there are still worth-while technological possibilities for the type II superconductors. The "*Technology Roadmap for the Superconducting Electronics Industry*" should prove useful in assessing the directions to take for the full commercial exploitation of these materials.

It is still too early to foresee the extent to which the high temperature super-conductors will be exploited and whether a highly profitable industry will emerge on the back of this revolutionizing technology. However, the prognosis seems good; products are already on the market, others are imminent. Similarly, it is not at all certain as to how high in temperature superconductivity can be observed and whether the dream of a room temperature superconductor will ever be reached. From a thermodynamic viewpoint there is nothing significant about room temperature. In the final analysis, achieving room temperature superconductivity will of course depend upon the laws of nature. Richard Feynman succinctly put this very obvious, but frequently forgotten fact, in interviews connected with his highly objective analysis of the Challenger space shuttle disaster of 1985. It is appropriate that we should conclude this book by referring to Richard Feynman, one of the great scientific figures of the second half of the twentieth century. Perhaps it is a measure of the truly complex nature of superconductivity that he spent several years at the peak of his career investigating the subject but was unable to come up with an explanation. In drawing attention to the simple but in the end fatal flaw responsible for the Challenger catastrophe he pointed out that "nature cannot be fooled".

References

Prologue

The following are some introductory books and articles on the subject of superconductivity:

Books

Schechter, B. The Path of No Resistance: The story of the Revolution in Superconductivity. Simon and Schuster (1989).
Vidali, G. Superconductivity: The Next Revolution. Cambridge University Press (1993).

Articles

Campbell, P. A Superconductivity Primer. *Nature* **330**, 21–4 (1987).
Chu, C.W. High Temperature Superconductors. *Scientific American* **273**, 128–31 (September 1995).
Day, P. Superconductors: Past, Present and Future. *Proc. Roy. Inst.* **65**, 29–46 (1994).
Ford, P.J. Superconductivity comes in from the cold. *South Afr. J. Sci.* **84**, 87–91 (1988).
Ford, P.J. and Saunders, G.A. High-Temperature Superconductivity – ten years on. *Contemporary Physics* **38**, 63–81 (1997).
Gough, C.E. High Temperature Superconductors Take Off. *Phys. Edu.* **33**, 38–46 (1998).
Gough, C.E. Challenges of High-T_c, *Phys. World* **4**, 26–30 (December 1991).
Kenward, M. The Heat is on for Superconductors. *New Scientist* **114**, No. 1559, 46–51 (7 May 1987).
Lemonick, M.D. Superconductors! *Time Magazine* 38–45 (11 May 1987).
Sang, D. Superconductivity. *New Scientist* **153**, No. 2065 (Inside Science No. 97) (18 January 1997).

Nobel Prizes

Gura, T. Eyes on the Prize. *Nature* **433**, 560–4 (11 October 2001).
Stone, R. At 100, Alfred Nobel's Legacy Retains its Luster. *Science* **294**, 288–91 (12 October 2001).

Many aspects of this present book have been discussed in far greater detail in the comprehensive two-volume set:

Handbook of Superconducting Materials. D. Cardwell and D. Ginley (ed.) 2174 pp. *Institute of Physics Publishing* (2003).
Volume 1: Superconductivity, Materials and Processes.
Volume 2: Characterization, Applications and Cryogenics.

Chapter 1

Books

Casimir, H.J. Haphazard Reality: Half a Century of Science. Harper and Row, New York (1983).
Dahl, P.P. Superconductivity – Its Historical Roots and Development from Mercury to the Ceramic Oxides. American Institute of Physics, New York (1992).
Mendelssohn, K. The Quest for Absolute Zero – The Meaning of Low Temperature Physics (2nd edition). Taylor and Francis (1977).

Articles

Lowering temperature in quest for its absolute zero

Ford, P.J. Towards the Absolute Zero: The Early History of Low Temperatures. *South African Journal of Science* **77**, 244–8 (1981).
Gorter, C.J. and Taconis, K.W. The Kamerlingh Onnes Laboratory. *Cryogenics* **4**, 345–53 (1964).
Kurti, N. From Cailletet and Pictet to Microkelvin. *Cryogenics* **18**, 451–58 (1978).
Kurti, N. From the First Mist of Liquid Oxygen to Nuclear Ordering: Anecdotes from the History of Refrigeration. *Physica* **109–110B**, 1737–52 (1982).

The electrical resistance of metals and the discovery of superconductivity

de Bruyn Ouboter, R. Heike Kamerlingh Onnes's Discovery of Superconductivity. *Scientific American* **276**, 84–89 (March 1997).
De Nobel, J. The Discovery of Superconductivity. *Physics Today* **49**, 40–2 (September 1996).
Meije, P.H.E. Kamerlingh Onnes and the Discovery of Superconductivity. *Am. J. Phys.* **62**, 1105–8 (1994).

Chapter 2

Books

There are numerous books devoted to the topics of superconductivity discussed in this chapter. Two excellent ones are:
Buckel, W. Superconductivity – Fundamentals and Applications. VCH Weinheim (1991).
Tinkham, M. Introduction to Superconductivity 2nd ed. McGraw-Hill Inc. New York (1996).
Likewise books on solid state physics invariably contain a chapter on superconductivity. See for example:
Kittel, C. Introduction to Solid State Physics, 7th ed. Wiley (1996).
The following contain material relevant to this chapter
Dahl, P.P. Superconductivity – Its Historical Roots and Development from Mercury to the Ceramic Oxides. American Institute of Physics, New York (1992).
Hoddeson, L., Braun, E., Teichman, J. and Weart, S. Out of the Crystal Maze – Chapters from the History of Solid State Physics. Oxford University Press (1992).

Early years

de Bruyn Ouboter, R. Superconductivity: Discoveries during the Early Years of Low Temperature Research at Leiden 1908–1914. *IEEE Transactions on Magnetics,*Vol. Mag.-23, 355–70 (March 1987).
Pippard, A.B. Early Superconductivity Research (Except Leiden). *IEEE Transactions on Magnetics,*Vol. Mag.-23, 371–75 (March 1987).
Shoenberg, D. Superconducting Colloidal Mercury. *Nature* 143, 434–5, (1939).
Thomas, H. Some Remarks on the History of Superconductivity "Earlier and Recent Aspects of Superconductivity" eds. Bednorz, J.G. and Müller, K.A. Springer Series in Solid State Sciences 90, Springer-Verlag p. 2–44, (1990).

A-15 Materials

Hulm, J.K. and Matthias, B.T. High-Field, High-Current Superconductors. *Science* **208**, 881–87 (1980).
Hulm, J.K., Kunzler, J.E. and Matthias, B.T. The Road to Superconducting Materials. *Physics Today* **35**, 34–43 (January 1982).
Matthias, B.T. The Search for High Temperature Superconductors. *Physics Today* **24**, 23–28 (August 1971).
Muller, J., A15-Type Superconductors. *Rep. Prog. Phys.* **43**, 641–87 (1980).
Rivlin, V. and Dew-Hughes, D. Getting Warmer in the Superconductivity Game. *New Scientist* **67**, 261–4 (July 31, 1975).
Testardi, L.R. Elastic Behaviour and Structural Instability of High Temperature A-15 Structure Superconductors. "Physical Acoustics" ed. W.P. Mason and R.N. Thurston **10**, pp. 193–296 (1973).
Testardi, L.R. Structural Instability and Superconductivity in A-15 Compounds. *Rev. Mod. Phys.* **47**, 637–48 (1975).
Testardi, L.R. Structural Instability of High Temperature A-15 Superconductors. "Physical Acoustics" ed. W.P. Mason and R.N. Thurston **13**, pp 29–47 (1977).

Other articles

Bardeen, J. Excitonic Superconductivity. *J. Less Common Metals* **62**, 447–50 (1978).
Bardeen, J., Cooper, L.N. and Schrieffer, J.R. Microscopic Theory of Superconductivity. *Phys. Rev.* **106**, 162–64 (1957).
Bardeen, J., Cooper, L.N. and Schrieffer, J.R. Theory of Superconductivity. *Phys. Rev.* **108**, 1175–1204 (1957).
Beasley, M.R. and Geballe, T.H. Superconducting Materials. *Physics Today* **37**, 60–8 (October 1984).
Brandt, N.B. and Ginsburg, V.I. Superconductivity at High Pressure. *Scientific American* **224**, 83–91 (April 1971).
Bromberg, J.L. Experiment vis-à-vis Theory in Superconductivity Research: the Case of Bernd Matthias. Physics, Philosophy and the Scientific Community ed. Gavroglu, K. *et al.* Kluwer Academic Publishers, Boston (1995).
Holton, G., Chang, H. and Jurkowitz, E. How a Scientific Discovery is Made: A Case History. *American Scientist* **84**, 364–75 (1996).
König, R., Schindler, A. and Hermannsdörfer, T. Superconductivity of Compacted Platinum Powder at Very Low Temperatures. *Phys. Rev. Lett.* **82**, 4528–31 (1999).
Matthias, B.T. High Temperature Superconductivity? *Comments on Solid State Physics* **3**, 93–6 (1970).
National Research Council. Physics through the 1990s. Condensed-Matter Physics, p. 189, National Academy Press (1986).

Schrieffer, J.R and Tinkham, M., Superconductivity. *Rev. Mod. Phys.* **71**, S 8313–17 (1999).

Shimitzu, K., Ishikawa, K., Takao, D., Yagi, T. and Amaya, K. Superconductivity in compressed lithium at 20K. *Nature* **419**, 597–9 (October 10th, 2002).

Struzhkin, V., Eremets, M., Gan, W., Mao, H. and Hemley, R. Superconductivity in Dense Lithium. *Science* **298**, 1213–5 (November 8th, 2002).

Biography

Abrikosov, A.A. My Years with Landau. *Physics Today* **26**, 56–60 (January 1973).

Ford, P.J. Peter Kapitza 1894–1984. *South African Journal of Science* **80**, 253–6 (1984).

Goodstein, D. and Goodstein, J. Richard Feynman and the History of Superconductivity. "History of Original Ideas and Basic Discoveries in Particle Physics" ed. Newman, H.B. and Ypsilantis, T. pp. 773–791. Plenum Press, New York (1996).

Hoddeson, L. and Daitch, V. The Life and Science of John Bardeen. Joseph Henry Press (2002).

Kojenikov, A. Lev Landau: Physicist and Revolutionary. *Physics World* **15**, 35–39 (June 2002).

Livanova, A., Landau: A Great Physicist and Teacher. Pergamon Press, Oxford (1978).

Pines, D. Richard Feynman and Condensed Matter Physics. *Physics Today* **42**, 61–66 (February 1989).

Schrieffer, J.R. John Bardeen and the Theory of Superconductivity. *Physics Today* **45**, 46–53 (April 1992).

Shoenberg, D. Forty Odd Years in the Cold. *Physics Bulletin* **29**, 16–19 (January 1978).

Shoenberg, D. Heinz London 1907–1970, Biographical Memoirs of Fellows of the Royal Society **17**, 441–61 (1971).

Shoenberg, D. Kurt Alfred Georg Mendelssohn 1906–1980 Biographical Memoirs of Fellows of the Royal Society **29**, 361–398 (1983).

Shoenberg, D. Piotr Leonidovich Kapitza 1894–1984 Biographical Memoirs of Fellows of the Royal Society **31**, 327–374 (1985).

Chapter 3

Early work on oxide superconductors

Johnston, D.C, Prakash, H., Zachariasen, W.H. and Viswanathan, R. High Temperature Superconductivity in the Li–Ti–O Ternary System. *Mat. Res. Bull.* **8**, 777–84 (1973).

Sleight, A.W., Gillson, J.L. and Bierstedt, F.E. High Temperature Superconductivity in the $BaPb_{1-x}Bi_xO_3$ System. *Sol. State Commun.* **17**, 27–8 (1975).

The breakthrough

Bednorz and Müller wrote several review articles describing their work. Their 1987 Nobel Prize lectures published in Reviews of Modern Physics the following year are an excellent summary of the state of the subject at that time.

Bednorz, J.G. and Müller, K.A. Perovskite-type Oxides – The New Approach to High-T_c Superconductivity. *Rev. Mod. Phys.* **60**, 585–600 (1988).

Müller, K.A. and Bednorz, J.G. The Discovery of a Class of High-Temperature Superconductors. *Science* **237**, 1133–39 (1987).

Müller, K.A. The Development of Superconductivity Research in Oxides. "Proceedings of the 10th Anniversary HTS Workshop on Physics, Materials and Applications" ed. Batlogg, B., Chu, C.W., Gubser, D.U, and Müller, K.A. pp. 3–16, World Scientific Publishing Co. Singapore (1997).

Paul Chu has also written interesting review articles:

Chu, C.W. High Temperature Superconductivity. "History of Original Ideas and Basic Discoveries in Particle Physics" ed. Newman, H.B. and Ypsilantis, T. pp. 793–836. Plenum Press, New York (1996).

Chu, C.W. Superconductivity Above 90K and Beyond. "Proceedings of the 10th Anniversary HTS Workshop on Physics, Materials and Applications" ed. Batlogg, B., Chu, C.W., Gubser, D.U, and Müller, K.A. pp. 17–31. World Scientific Publishing Co. Singapore (1997).

Lanthanum–barium–copper–oxide

Bednorz, J.G. and Müller, K.A. Possible High T_c Superconductivity in the Ba–La–Cu–O System. *Z. Phys. B* **64**, 189–93 (1986).

Bednorz, J.G., Takashige M., and Müller, K.A. Susceptibility Measurements Support High-T_c Superconductivity in the Ba–La–Cu–O System. *Europhys. Lett.* **3**, 379–85 (1987).

Bednorz, J.G., Müller, K.A. and Takashige, M. Superconductivity in Alkaline Earth-Substituted La_2CuO_{4-y}. *Science* **236**, 73–5 (1987).

Cava, R.J., van Dover, R.B., Batlogg, B. and Rietmann, E.A. Bulk Superconductivity at 36K in $La_{1.8}Sr_{0.2}CuO_4$. *Phys. Rev. Lett.* **58**, 408–10 (1987).

Chu, C.W., Hor, P.H., Meng, R.L., Gao, L., Huang, Z.J. and Wang, Y.Q. Evidence for Superconductivity above 40K in the La–Ba–Cu–O Compound System. *Phys. Rev. Lett.* **58**, 405–7 (1987).

Chu, C.W., Hor, P.H., Meng, R.L., Gao, L. and Huang, Z.J. Superconductivity at 52.5K in the Lanthanum–Barium–Copper–Oxide System. *Science* **235**, 567–9 (1987).

Chu, C.W., Gao, L., Chen, F., Huang, Z.J., Meng, R.L. and Xue, Y.Y. Superconductivity above 150K in $HgBa_2Ca_2Cu_3O_{8+\delta}$. *Nature* **365**, 323–25 (1993).

Michel, C., Er-Rakho, L. and Raveau, B. The Oxygen Defect Perovskite $BaLa_4Cu_5O_{13.4}$, A Metallic Conductor. *Mat. Res. Bull.* **20**, 667–71 (1985).

Takagi, H., Uchida, S., Kitazawa, K. and Tanaka, S. High T_c Superconductivity of La–Ba–Cu Oxides II – Specification of the Superconducting Phase. *Jpn. Journ. Appl. Phys. (Letters)* **26**, L123–4 (1987).

Takagi, H., Uchida, S., Kitazawa, K. and Tanaka, S. The Influence of Oxygen Deficiency on the Electrical Resistivity of High T_c Superconducting Oxides $(La.Ba)_2CuO_{4-y}$. *Jpn. Journ. Appl. Phys. (Letters)* **26**, L218–9 (1987).

Uchida, S., Takagi, H., Kitazawa, K. and Tanaka, S. High T_c Superconductivity of La–Ba–Cu Oxides. *Jpn. Journ. Appl. Phys. (Letters)* **26**, L1–2 (1987).

Uchida, S., Takagi, H., Kitazawa, K. and Tanaka, S. High T_c Superconductivity of La–Ba–Cu Oxides III – Electrical Resistivity Measurements. *Jpn. Journ. Appl. Phys. (Letters)* **26**, L151–2 (1987).

Uchida, S., Takagi, H., Tanaka, S., Nakao, K., Miura, N., Kishio, K., Kitazawa, K. and Fueki, K. High T_c Superconductivity of La–Ba(Sr)–Cu Oxides IV – Critical Magnetic Fields. *Jpn. Journ. Appl. Phys. (Letters)* **26**, L196–7 (1987).

Yttrium–barium –copper–oxide

Wu, M.K., Ashburn, J.R., Torng, C.J., Hor, P.H., Meng, R.L., Gao, L., Huang, Z.J., Wang, Y.Q. and Chu, C.W. Superconductivity at 93K in a New Mixed Phase Y–Ba–Cu–O Compound System at Ambient Pressure. *Phys. Rev. Lett.* **58**, 908–10 (1987).

Hor, P.H., Gao, L., Meng, R.L., Huang, Z.J., Wang, Y.Q., Forster, K., Vassilious, J., Chu, C.W., Wu, M.K., Ashburn, J.R. and Torng, C.J. High-Pressure Study of the New Y-Ba-Cu-O Superconducting Compound System. *Phys. Rev. Lett.* **58**, 911-12 (1987).

Hor, P.H., Meng, R.L., Wang, Y.Q., Gao, L., Huang, Z.J., Bechtold, J., Forster, K. and Chu, C.W. Superconductivity above 90K in the Square-Planar-Compound System $ABa_2Cu_3O_{6+x}$. *Phys. Rev. Lett.* **58**, 1891-4 (1987).

Impact of high temperature superconductors

Felt, U. and Nowotny, H. Striking Gold in the 1990s: The Discovery of High Temperature Superconductivity and its Impact on the Science System. *Science, Technology and Human Values* **17**, 506-531 (1992).

Nowotny, H. and Felt, U. After the Breakthrough. Cambridge University Press (1997).

Nowotny, H. Millennium Essay: Innovation Machine on the Boil. *Nature* **401**, 859 (1999).

Hazen, R.M. The Breakthrough, The Race for the Superconductor. Summit Books (1988).

Hazen, R.M., Perovskites. *Scientific American* **258**, 52-61 (June 1988).

Holton, G., Chang, H. and Jurkowitz, E. How a Scientific Discovery is Made: a Case History. *American Scientist* **84**, 364-75 (1996).

Pool, R. Superconductor Credits Bypass Alabama. *Science* **241**, 655-7 (1988).

Bismuth, thallium and mercury superconductors

Chu, C.W., Gao, L., Chen, F., Huang, Z.J., Meng, R.L. and Xue, Y.Y. Superconductivity above 150K in $HgBa_2Ca_2Cu_3O_{8+\delta}$ at High Pressures. *Nature* **365**, 323-5 (1993).

Hazen, R.M., Prewitt, C.T., Angel, R.J., Ross, N.L., Finger, W., Hadidiacos, C.G., Veblen, D.R., Heaney, P.J. Hor, P.H., Meng, R.L, Sun, Y.Y., Wang, Y.Q., Xue, Y.Y., Huang, Z.J. Gao, L., Bechtold, J., and Chu, C.W. Superconductivity in the High T_c Bi-Ca-Sr-Cu-O System: Phase Identification. *Phys. Rev. Lett.* **60**, 1174-77 (1988a).

Hazen, R.M., Finger, L.W., Angel, R.J., Prewitt, C.T., Ross, N.L. Hadidiacos, C.G., Heaney, P.J. Veblen, D.R., Sheng, Z.Z., El Ali, A. and Hermann A.M. 100-K Superconductivity Phases in the Tl-Ca-Ba-Cu-O System. *Phys. Rev. Lett.* **60**, 1657-60 (1988b).

Maeda, H., Tanaka, Y., Fukutomi, M. and Asano.T. A New High T_c Oxide Superconductor without a Rare Earth Element. *Jpn. Journ. Appl. Phys.* **27**, L209-10 (1988).

Schilling, A., Cantoni, M., Guo, J.D. and Ott, H.R. Superconductivity above 130K in the Hg-Ba-Ca-Cu-O System. *Nature* **363**, 56-58 (1993).

Sheng, Z.Z. and Hermann, A.M. Superconductivity in the Rare-Earth Free Tl-Ba-Cu-O System above Liquid Nitrogen Temperature. *Nature* **332**, 55-8 and 138 (1988).

Sheng, Z.Z. and Hermann, A.M. Bulk Superconductivity at 120K in the Tl-Ca/Ba-Cu-O System. *Nature* **332**, 138-9 (1988).

Chapter 4

There is a huge literature on the structure and properties of the high temperature super-conductors. Two excellent reference sources are the series "Studies of High Temperature Superconductors" ed. A. Narliker. Nova Science Publishers Inc. There are some 56 volumes in this series to date and they deal with experiment, theory and applications. The second is "Physical Properties of High Temperature Superconductors" ed. D.M. Ginsburg; World Scientific. This has five volumes so far in theory and experiment.

The following contain excellent discussions on the structure of high temperature super-conductors:

Aranda, M.A.G. Crystal Structure of Copper-Based High-T_c Superconductors. *Adv. Mater.* **6**, 905-21 (1994).

Cava, R.J., Superconductors Beyond 1–2–3. *Scientific American* **263**, 24–31 (August 1990).
Grant, P.M., High Temperature Superconductivity: Four Years since Bednorz and Müller, *Adv. Mater.* **2**, 232–53 (1990).
Hazen, R.M. Crystal Structure of High Temperature Superconductors. Physical Properties of High Temperature Superconductors II, ed D.M. Ginsburg; World Scientific, pp. 121–98 (1990).
Longo, J.M. and Raccah, P.M. The structure of La_2CuO_4 and $LaSrVO_4$. *J. Solid State Chemistry* **6**, 526–31 (1973).
Raveau, B., Michel, C. and Hervieu, M. Structural Chemistry of High T_c Superconductors. Studies of High Temperature Superconductors 2, Nova Science Publishers Inc. ed. A. Narliker, 1–27 (1989).
Sleight, A.W. Chemistry of High-Temperature Superconductors. *Science* **242**, 1519–27 (1988).

Other articles

Antipov, E.V., Abakumov, A.M. and Putilin, S.N. Chemistry and Structure of Hg-based Superconducting Mixed Oxides. *Supercond. Sci. Technol.* **35**, R31–49 (2002).
Batlogg, B. Selected Experiments on High T_c Cuprates. The Los Alamos Symposium-1989 High Temperature Superconductivity Proceedings ed. K.S. Bedell, D. Coffey, D.E. Meltzer, D. Pines and J.R. Schrieffer. Addison-Wesley Publishers 37–93 (1990).
Batlogg, B. Physical Properties of High-T_c Superconductors. *Physics Today* **44**, 44–50 (June 1991).
Beyers, R. and Shaw, T.M. The Structure of $Y_1Ba_2Cu_3O_{7-x}$ and its Derivatives. Solid State Physics **42**, ed. Ehrenreich, H. and Turnbull, D. 135–212 (1989).
Beyers, R. and Ann, B.T. Thermodynamic Considerations in Superconducting Oxides, *Annual Review of Materials Science* **21**, 335–435 (1991).
Bordet, P. Capponi, J.J., Chaillout, C, Chenavas, J., Hewat, A.W., Hewat, E.A., Hodeau, J.L. and Marezio, M. A Review of the Preparation and Structure of Bi–Sr–Ca–Cu–O Superconductors and Pb-Substituted Phases. Studies of High Temperature Superconductors 2, Nova Science Publishers Inc. ed. A. Narliker 171–98 (1991).
Bryntse, I. A Review of the Synthesis and Properties of Hg-Containing Superconductors, Focused on Bulk Materials & Thin Films. "Studies of High Temperature Superconductors **23**, 1–26" ed. A. Narliker. Nova Science Publishers Inc. (1997).

Both vol. **23–24** of this series are devoted entirely to Hg-based high temperature superconductors.

Cava, R.J. Structural Chemistry and the Local Charge Picture of Copper Oxide Superconductors. *Science* **247**, 656–62 (1990).
Cava, R.J., Hewat, A.W., Hewat, E.A., Batlogg, B., Marezio, M., Rabe, K.M., Krajewski, J.J., Peck, W.F., and Rupp, L.W. Structural Anomalies, Oxygen Ordering and Superconductivity in Oxygen Deficient $Ba_2YCu_5O_x$. *Physica C* **165**,419–33 (1990).
Fisk, Z. and Sarrao, J.L. The New Generation High-Temperature Superconductors. *Annual Review of Materials Science* **27**, 35–67 (1997).
Geballe, T.H. and Hulm, J.K. Superconductivity – The State that came in from the Cold. *Science* **239**, 367–75 (1988).
Grant, P.M. Do-it-yourself Superconductors, *New Scientist* **115**, 36–9 (July 30 1987).
Jorgensen, J.D. Defects and Superconductivity in the Copper Oxides. *Physics Today* **44**, 34–40 (June 1991).
Malozemoff, A.P. and Grant, P.M. High Temperature Superconductivity Research at the IBM Thomas J. Watson and Almaden Research Centers. *Z. Phys. B* **67**, 275–83 (1987).
Sleight, A.W. Synthesis of Oxide Superconductors. *Physics Today* **44**, 24–30 (June 1991).
Ziman, J.D. The Ordinary Transport Properties of the Noble Metals. *Adv. in Phys.* **10**, 1–56 (1961).

Chapter 5

There is a vast literature on atomic theory and quantum mechanics and only a few references will be cited.

Books

Blin-Stoyle, R. Eureka! – Physics of Particles, Matter and the Universe. Institute of Physics Publishing, Bristol (1997).
Born, M. Atomic Physics. Blackie Press 7th ed. 1962.
Landsberg, P.T. Seeking Ultimates. Institute of Physics Publishing, Bristol (2000).

Articles

Bohr, N. The Structure of the Atom. Nobel Prize for Physics lecture 1922. Nobel Lectures Physics 1922–41, pp. 7–43. Elsevier Publishing Company, Amsterdam (1965).
Ford, PJ. Niels Bohr 1885–1961. *S. Afr. J. Science* **82**, 179–189 (1986); ibid **83**, 15–21 (1987); ibid **84**, 170–9 (1988).
Goudsmit, S.A. Fifty Years of Spin – It Might as well Spin. *Physics Today* 40–43 (June 1976).
Pais, A. Introducing atoms and their nuclei. Chapter 2, 43–142 Twentieth Century Physics ed. Brown, L.M., Pais, A. and Pippard, A.B. Institute of Physic Publishing, Bristol and Philadelphia 1995.
Rechenburg, H. Atoms and Quanta. Chapter 3, 143–248 Twentieth Century Physics ed. Brown, L.M., Pais. A. and Pippard, A.B. Institute of Physic Publishing, Bristol and Philadelphia 1995.
Uhlenbeck, G.E. Fifty Years of Spin – Personal Reminiscences. *Physics Today* 43–8 (June 1976).

All the Standard books on Condensed Matter Physics would have a chapter devoted to electrons in metals. The following are some useful references:

Book

Ziman, J.M. Electrons in Metals – A Short Guide to the Fermi Surface. Taylor & Francis Ltd, London 1970.

Articles

Flouquet, J. and Buzdin, A. Ferromagnetic Superconductors. *Physics World* **15**, 41–6 (2002).
Law, S. Electrons Switch on to Heavy Metal. *New Scientist* **126**, 57–60 (19 May 1990).
Pippard, A.B. Electrons in Solids. Chapter **17**, 1279–1383, Vol. III Twentieth Century Physics ed. Brown, L.M., Pais, A. and Pippard, A.B. Institute of Physics Publishing, Bristol and Philadelphia (1995).

Fermi surfaces

Mackintosh, A.R. The Fermi Surface of Metals. *Scientific American* **209**, 110–20 (July 1963).

Pickett, W.E., Krakauer, H., Cohen, R.E. and Singh, D.J. Fermi Surfaces, Fermi Liquids and High-Temperature Superconductors. *Science* **255**, 46–54 (1992).

Schoenberg, D. Electrons at the Fermi Surface. Chapter **5**, 109–140. Solid State Science; Past, Present and Predicted ed. Weaire, D.L. and Windsor, C.G. Institute of Physics Publishing (1987).

Scott, G.B. and Springford, M. The Fermi Surface of Niobium. *Proc. Roy. Soc. A* **320**, 115–130 (1970).

Chapter 6

What causes superconductivity?

Nobel Prize Lectures in Physics 1972:

Bardeen, J. Electron–Phonon Interactions and Superconductivity. *Physics Today* **26**, 41–6 (July 1973).

Cooper, L.N. Microscopic Interference in the Theory of Superconductivity. *Physics Today* **26**, 31–39 (July 1973).

Schrieffer, J.R. Macroscopic Quantum Phenomena from Pairing in Superconductors. *Physics Today* **26**, 23–8 (July 1973).

Other articles

Bardeen, J. Superconductivity and other Macroscopic Quantum Phenomena. *Physics Today* **43**, 25–31 (December 1990).

Bardeen, J. and Schrieffer, J.R. Recent Developments in Superconductivity Progress Low Temperature Physics, Vol. III, 170–287, ed. by C.J. Gorter (Amsterdam: North-Holland Publishing Co.) (1961).

Bardeen, J., Cooper, L.N. and Schrieffer, J.R. Microscopic Theory of Superconductivity. *Phys. Rev.* **108**, 1175–1204 (1957).

Cooper, L.N. Theory of Superconductivity. *American Journal of Physics* **28**, 91–101 (1960).

Douglas, D.H. and Falicov, L.M. Progress in Low Temperature Physics, ed. C.J. Gorter (Amsterdam: North-Holland Publishing Co.) Vol. IV, p. 97 (1964).

Geballe, T.H. This Golden Age of Solid-State Physics. *Physics Today* **34**, 132–143 (1981)

Giaver, I. Electron Tunneling and Superconductivity. *Nobel Lectures Physics* 1971–80, ed. Stig Lundqvist, pp. 137–153, World Scientific (Singapore) (1992).

Ginzburg, V.L. John Bardeen and the Theory of Superconductivity. The Physics of a Lifetime. 451–56. Springer (Berlin) (2001).

Saunders, G.A. The Electron Pair Theory of Superconductivity. *Contemp. Phys.* **7**, 192–209 (1966).

Schrieffer, J.R. Theory of Superconductivity (Benjamin) (1964).

Waldram, J.R. Superconductivity of Metals and Cuprates. Chapters 6–10. Institute of Physics Publishing (Bristol and Philadelphia) (1996).

Ultrasonic investigations

Saunders, G.A. and Lawson, A.W. Ultrasonic Determination of the Superconducting Energy Gap in Thallium, *Phys. Rev.* **135A**, 1161–5 (1964).

Saunders, G.A. and Lawson, A.W. Ultrasonic Determination of the Superconducting Energy Gap in In_2Bi. *J. Applied Physics* **35**, 3322–4 (1964).

Superconductivity in magnesium-diboride

Finnemore, D.K., Ostenson, J.E., Bud'ko, S.L., Lapertot, G. and Canfield P.C. Thermodynamic and Transport Properties of $Mg^{10}B_2$. *Phys. Rev. Letters* **86**, 1877–80 (2001).

Canfield, P.C. and Bud'ko, S.L. Magnesium diboride: One year on. *Physics World* **15**, 29–34 (January 2002).

Canfield, P.C. and Crabtree, G.W. Magnesium Diboride: Better Late than Never. *Physics Today* **56**, 34–40 (March 2003).

Gough, C.E. New Metallic Superconductor Makes an Immediate Impact. *Physics World* **14**, 21–2 (May 2001).

Nagamatsu, J., Nakagawa, N., Muranaka, T. and Akimitsu, J. Superconductivity at 39K in magnesium diboride. *Nature* **410**, 63–4 (2001).

Yildirim, T. The Surprising Superconductor. *Materials Today* 40–45 (April 2002).

Chapter 7

Flux quantization

Deaver, B.S. and Fairbank, W.M. Experimental evidence for quantised flux in superconducting cylinders. *Phys. Rev. Lett.* **7**, 43–6 (1961).

Doll, R. and Nabauer, M. Experimental proof of magnetic flux quantization in a superconducting ring. *Phys. Rev. Lett.* **7**, 51–2 (1961).

Gough, C.E., Colclough, M.S., Forgan, E.M., Jordan, R.G., Keene, M., Muirhead, C.M., Rae, A.I.M., Thomas, N., Abell, J.S. and Sutton, S. Flux quantization in a high-T_c superconductor. *Nature* **326**, 855 (1987).

Gough, C.E. Flux Quantisation and Quantum Coherence in Conventional and HTC Superconductors and their Application to SQUID Magnetometry. Earlier and Recent Aspects of Superconductivity eds. Bednorz, J.G. and Müller, K.A.; Springer Series in Solid State Sciences 90 Springer-Verlag pp. 141–62 (1990).

Gough, C.E. Flux Quantisation, Superconducting Mixed States, the Josephson Effect and Squid Devices. High Temperature Superconductivity ed. D.P. Tunstall and W. Barford. Adam Hilger pp. 37–58 (1991).

Josephson effects

Early history

Anderson, P.W. How Josephson Discovered his Effect. *Physics Today* **23**, 23–9 (November 1970).

Anderson, P.W. and Rowell, J.M. Probable Observation of the Josephson Tunnelling Effect. *Phys. Rev. Lett.* **10**, 230–2 (1963).

Josephson, B.D. Possible New Effects in Superconductive Tunnelling. *Phys. Lett.* **1**, 251–3 (1962).

Josephson, B.D. The Discovery of Tunnelling Supercurrents. *Rev. Mod. Phys.* **46**, 251 (1974).

McDonald, D.G. The Nobel Laureate versus the Graduate Student. *Phys. Today* **54**, 46–51 (July 2001).

Rowell, J.M. Magnetic field dependence of the Josephson tunnel current effects. *Phys. Rev. Lett.* **11**, 200–2 (1963).

Rowell, J.M. Superconducting Tunneling Spectroscopy and the Observation of the Josephson Effect. *IEEE Trans. Mag.* **23**, 380–9 (1987).
Petley, B.W. The Josephson Effects. *Contemp. Phys.* **10**, 139–58 (1969).

Chapter 8

An excellent and comprehensive reference source on material contained in this and the following chapter is:

"Handbook of Applied Superconductivity" ed. B. Seeber.
Volume 1: Fundamental Theory, Basic Hardware and Low Temperature Science and Technology.
Volume 2: Applications.
Institute of Physics Publishing, Bristol (1998).

Some straightforward articles about the potential applications of superconductors both large scale and in superconducting electronics include:

Burgoyne, J.W. Superconductivity Leaves the Lab. *Physics World* **13**, 23–4 (October 2000).
Kenward, M. Superconductors Get Ready to Transform Industry. *Physics World* **9**, 29–31 (June 1996).
Lubkin, G. Applications of High-Temperature Superconductors Approach the Marketplace. *Physics Today* **48**, 20–3 (March 1995).
Rosner, C.H. Superconductivity: Star Technology for the 21st Century. *IEEE Trans. on Applied Superconductivity* **11**, 39–48 (2001).
Silver, T.M., Dou, S.K. and Jin, J.X. Applications of High Temperature Superconductors. *Europhysics News* **32**, 82–6 (2001).
Tallon, J. Industry Warms to Superconductors. *Physics World* **13**, 27–31 (March 2000).

Superconducting cables and power transmission

An introductory series of articles on several aspects of power distribution, with the title "Superpower" edited by U. Balachandran appeared in the journal *IEEE Spectrum* for July 1997. The articles are:

Hull, J.R. Flywheels on a Roll. 20–5.
Leung, E., Surge Protection for Power Grids. 26–30.
Rahman, M.M. and Nassi, M. High-Capacity Cable's Role in once and Future Grids. 31–5.
Blaugher, R.D. Low-Calorie, High-Energy Generators and Motors. 36–42.
Mehta, S.P., Aversa, N. and Walker, M.S. Transforming Transformers. 43–9.

Other references on superconducting cables and power transmission

Beales, T.P. Towards a High Temperature Superconducting Power Transmission Cable. *App. Phys. Comm.* **12**, 205–20 (1993).
Grant, P.M. Superconductivity and Electric Power: Promises, Promises... Past, Present and Future. *IEEE Trans. on Applied Superconductivity* **7**, 112–133 (1997).
Hassenzahl, W.V. Superconductivity, an Enabling Technology for the 21st Century Power Systems. *IEEE Trans. on Applied Superconductivity* **11**, 1447–53 (2001).
Hull, J.R. Applications of High-temperature Superconductors in Power Technology. *Reports of Progress in Physics* **66**, 1865-86 (2003).
Komarek, P. Superconductivity in Technology. *Contemporary Physics* **17**, 355–86 (1976).
Larbalestier, D. The Road to Conductors: 10 Years do make a Difference! *IEEE Trans. on*

Applied Superconductivity **7**, 90–97 (1997).

Larbalestier, D., Gurevich, A., Feldmann, D. and Polyanskii, A. High-T_c Superconducting Materials for Electric Power Applications. *Nature* **414**, 368–77 (2001).

Leghissa, M., Rieger, J., Wiezoreck, J., and NeuMuller, H.-W. HTS Cables for Electric Power Transmission: Basic Properties – State of the Art – Prospects. *Advances in Solid State Physics* **38**, 551–64 (1998).

Lubkin, G. Power Applications of High Temperature Superconductors. *Physics Today* **49**, 48–51 (March 1996).

Marsh, G. Time Ripe for Superconductivity? *Materials Today* 46–50 (April 2002).

Sato, K.-I. Super Conductors. *Physics World* **5**, 37–40 (July 1992)

Magnetic fields, critical currents, flux pinning etc.

Bending, S.J. Local Magnetic Probes of Superconductors. *Adv. Phys.* **48**, 449–535 (1999).

Bishop, D.J., Gammel, P.J. and Huse, D.A. Resistance in High Temperature Superconductors. *Scientific American* **268**, 24—31 (February 1993).

Brandt, E.H. The Flux-Line Lattice in Superconductors. *Rep. Prog. Phys.* **58**, 1465–1594 (1995).

Essmann, U. and Träuble, H. The Magnetic Structure of Superconductors. *Scientific American* **224**, 74–84 (March 1971).

Huse, D.A., Matthew, P.A. and Fisher, D.S. Are Superconductors Really Superconducting? *Nature* **358**, 553–9 (1992).

Larbalestier, D., Fisk, G., Montgomery, B. and Hawksworth, D. High-field Superconductivity. *Physics Today* **39**, 24–33 (March 1986).

Larbalestier, D. Critical Currents and Magnet Applications of High-T_c Superconductors. *Physics Today* **44**, 74–82 (June 1991).

Raveau, B. Defects and Superconductivity in Layered Cuprates. *Physics Today* **45**, 53–8 (October 1992).

Zheng, H., Jiang, M., Niklova, R., Welp, U., Paulikas, A.P., Huang, Yi, Crabtree, G.W., Veal, B.W. and Claus, H. High Critical Current "Weld" Joints in Textured $YBa_2Cu_3O_x$. *Physica C* **322**, 1–8 (1999).

Applications of superconducting magnets

A vast amount of recent information on superconducting magnets can be found in the Proceedings of the Seventeenth International Conference on Magnet Technology held at CERN, Geneva in September 2001 and appearing in IEEE Transactions on Applied Superconductivity **12**, (2002).

Magnetic resonance imaging (MRI)

Oppelt, A. and Grandke, T. Magnetic Resonance Imaging. *Supercond. Sci. and Technol.* **6**, 381–95 (1993).

Guy, C.N. The Second Revolution in Medical Imaging. *Contemp. Phys.* **37**, 15–45 (1996).

Hounsfield, G.N. Computed Medical Imaging. *Science* **210**, 22–28 (1980).

Lauterbur, P.C. Image Formation by Induced Local Interactions: Examples Employing Nuclear Magnetic Resonance. *Nature* **242**, 190–1 (1973).

Nuclear and elementary particle physics

Geballe, T.H. and Rowell, J.M. Funding the SSC. *Science* **259**, 1237–8 (1993).

Glashow, S.L. and Lederman, L.M. The SSC. A Machine for the Nineties. *Physics Today* **38**, 28–37 (March 1985).

Kane, G.L. Physics Goals of the Superconducting SuperCollider (SSC). *Contemporary Physics* **34**, 69–78 (1993).

Kane, G.L. Supersymmetry – Squarks, Photinos, and the Unveiling of the Ultimate Laws of Nature. Perseus Publishing, Cambridge, Massachusetts (2000).

Ross, G.G. A "Theory of Everything"? *Contemporary Physics* **34**, 79–88 (1993).

Salam, A. Unification of Fundamental Forces. Cambridge University Press (1980).

Sutter, D.F. and Strauss, B.P. New Generation High Energy Physics Colliders: Technical Challenges and Prospects. *IEEE Transactions on Applied Superconductivity* **10**, 33–43 (2000).

LHC lattice magnets enter production. *CERN Courier* **41**,15–16 (June 2001).

LHC insertions: the key to CERN's new accelerator. *CERN Courier* **41**, 28–29 (October 2001).

Chapter 9

Chaloupka, H.J. High Temperature Superconductor Antennas: Utilisation of Low rf Losses and Non-linear Effects. *Journal of Superconductivity* **5**, 403–16 (1992).

Gallop, J. Microwave Applications of High-Temperature Superconductors. *Supercond. Sci. Technol.* A120–41 (1997).

Gough, C.E. Superconducting Antennas. Superconducting Technology – 10 Case Studies ed. K. Fossheim, World Scientific pp. 137–151 (1991).

Hilgenkamp, H. and Mannhart, J. Grain Boundaries in High-T_c Superconductors. *Reviews of Modern Physics* **74**, 485–549 (2002).

Hayakawa, H. Josephson Computer Technology. *Physics Today* **39**, 46–52 (March 1986).

Klein, N. High-frequency Applications of High-temperature Superconductor Thin Films. *Reports of Progress in Physics* **65**, 1387–1425 (2002).

Lancaster, M.J. Passive Microwave Device Applications of High Temperature Superconductors, Cambridge University Press (1996).

Mannhart, J. and Chaudhari, P. High-T_c Bicrystal Grain Boundaries. *Physics Today* **54**, 48–53 (November 2001).

Norton, D.P. Science and Technology of High Temperature Superconducting Films. *Annual Review of Materials Science* **67**, 299–347 (1998).

Phillips, J.M. Substrate Selection for High-Temperature Superconducting Thin Films. *Journ. of Applied Physics* **79**, 1829–48 (1996).

Richards, P.L. Analog Superconducting Electronics. *Physics Today* **39**, 54–62 (March 1986).

Richards, P.L. Bolometers for Infrared and Millimeter Waves. *J. Appl. Phys.* **76**, 1–24 (1994).

Simon, R. High-Temperature Superconductors for Microelectronics. *Solid State Technology* pp. 141–6 (September 1989).

Simon, R. High-T_c Thin Films and Electronic Devices. *Physics Today* **44**, 64–70 (June 1991).

Silver, A.H. Superconductivity in Electronics. *IEEE Transactions on Applied Superconductivity* **7**, 69–79 (1997).

Van Duzer, T. Superconductor Electronics, 1986–1996 Future. *IEEE Transactions on Applied Superconductivity* **7**, 98–110 (1997).

Wördenweber, R. Growth of High-T_c Films. *Supercond. Sci. Technol.* **12**, R86–102 (1999).

SQUIDS

Book

Gallop, J.C. SQUIDS, the Josephson Effects and Superconducting Electronics. Adam Hilger, Bristol (1990).

Articles

Clarke, J. Electronics with Superconducting Junctions. *Physics Today* **24**, 30–37 (August 1971).
Clarke, J. SQUIDs, Brains and Gravity Waves. *Physics Today* **39**, 36–44 ((March 1986).
Clarke, J. SQUID's: Theory and Practice. The New Superconducting Electronics ed. H. Weinstock and R.W. Ralston. pp. 123–142, Kluwer Academic Publishers (1993).
Clarke, J. SQUIDs. *Scientific American* **271**, 36–43 (August 1994).
Clarke, J. and Koch, R.H. The Impact of High-Temperature Superconductivity on SQUID Magnetometers. *Science* **242**, 217–23 (1988)
Clarke, J. A Superconducting Galvanometer Employing Josephson Tunnelling. *Phil. Mag.* **13**, 115–27 (1966).
Gallop, J.C. and Petley, B.W. SQUIDs and their applications. *J. Phys. E* **9**, 417–29 (1976).
Koelle, D., Kleiner, R., Ludwig, F., Dantsker, E. and Clarke, J. High-Transition-Temperature Superconducting Quantum Interference Devices. *Rev. Mod. Phys.* **71**, 631–86 (1999).
Spiller, T. and Clark, T. SQUIDs: macroscopic quantum objects. *New Scientist* **12**, 36–41 (4 December 1986).

SQUIDS and the brain

Hari, R. and Lounasmaa, O.V. Neuromagnetism: tracking the dynamics of the brain. *Physics World* **13**, 33–8 (May 2000).
Lounasmaa, O.V., Hamalainen, M., Hari, R. and Salmelin, R. Information processing in the human brain: Magnetocephalographic approach. *Proc. Nat. Acad. Sci. USA Science* **93**, 8809–15 (1996).
Ebert, T., Pantev, C., Wienbruch, C., Rockstroh, B. and Taub, E. Increased Cortical Representation of the Fingers of the Left Hand in String Players. *Science* **270**, 305–7 (1995).
Pizzella, V., Penna, S.D, Gratta, C.D. and Romani, G.L. SQUID systems for biomagnetic imaging. *Superconducting Science and Technology* **14**, R79–114 (2001).

SQUIDS and geophysics

Clarke, J. Geophysical Applications of SQUIDS. *IEEE Transactions on Magnetics* **19**, 288–94 (1983).

SQUIDS and non-destructive testing

Kirtley, J.R. and Wikswo, J.R. Scanning Squid Microscopy. *Annual Review of Materials Science* **29**, 117–48 (1999).

Chapter 10

Organic superconductors

The principal discoveries relating to possible and actual superconductivity in organic materials have been discussed in article in the "Search and Discovery" section of *Physics Today*. These refer to the original articles.

Superconducting fluctuations at 60K? *Physics Today* **26**, 17–19 (May 1973).

Attempts to Explain and Observe TCNQ Behaviour at 60K? *Physics Today* **26**, 17–19 (September 1973).

Organic Charge-Transfer Salt Shows Superconductivity. *Physics Today* **34**, 17–19 (February 1981).

The following are more comprehensive articles:

Bechgaard, K. and Jerome, D. Organic Superconductors. *Scientific American* **247**, 50–9 (July 1982).

Bechgaard, K. and Jerome, D. Organic Conductors and Organic Superconductors. *Physica Scripta* **T39**, 37–44 (1991).

Berlinsky, A.J. Organic Metals. *Contemp. Phys.* **17**, 331–54 (1976).

Chaikin, P.M. and Greene, R.L. Superconductivity and Magnetism in Organic Metals. *Physics Today* **39**, 24–32 (May 1986).

Friedel, J. The role of electron correlations in organic conductors. Organic Superconductivity "20th Anniversary" *J. de Physique IV*, **10**, 3–7 (2000).

Friedel, J. and Jerome, D. Organic Superconductors: the $(TMTSF)_2X$ Family. *Contemporary Physics* **23**, 583–624 (1982).

Little, W.A. Possibility of Synthesising an Organic Superconductor. *Phys. Rev.* **134A**, 1416–24 (1964).

Little, W.A. Superconductivity at Room Temperature. *Scientific American* **212**, 21–7 (February 1965)

Singleton, J. and Mielke, C. Superconductors go Organic. *Physics World* **15**, 35–9 (January 2002).

Singleton, J. and Mielke, C. Quasi-two-dimensional organic superconductors: a review. *Contemporary Physics* **43**, 63–96 (2002).

Fullerenes

There is a large literature on the fullerenes, which has had a considerable impact both on basic science and technology. This was recognized by the award of the 1996 Nobel Prize for Chemistry to Robert Curl, Harold Kroto and Richard Smalley. Their Nobel Prize lectures are reproduced in *Reviews of Modern Physics* for July 1997 and provide excellent accounts of the events leading to the discovery of the fullerenes.

Curl, R.F. Dawn of the Fullerenes: Experiment and Conjecture. *Rev. Mod. Phys.* **69**, 691–702 (1997).

Kroto, H. Symmetry, Space, Stars and C_{60}. *Rev. Mod. Phys.* **69**, 703–22 (1997).

Smalley, R.E. Discovering the Fullerenes. *Rev. Mod. Phys.* **69**, 723–30 (1997).

General articles on fullerenes

Curl, R.F. and Smalley, R.E. Probing C_{60}. *Science* **242**,1017–22 (1988).

Curl, R.F. and Smalley, R.E. Fullerenes. *Scientific American* **265**, 32–41 (October 1991).

Huffman, D.R. Solid C_{60}. *Physics Today* **44**, 22–29 (November 1991).

Kroto, H. Space, Stars, C_{60} and Soot. *Science* **242**, 1139–45 (1988).

Kroto, H.W. C_{60}: Buckminsterfullerene, the Celestial Sphere that fell to Earth. *Proceedings of the Royal Institution* **67**, 17–42 (1996).

Kroto, H.W., Allaf, A.W. and Balm, S.P. C_{60} Buckminsterfullerene. *Chemical Review* **91**, 1213–35 (1991).

Superconductivity in fullerenes

Gärtner, S. Superconductivity in Doped Fullerenes. Festkörprobleme **32**, 295–315 (1992).

Hebard, A.F. Superconductivity in Doped Fullerenes. *Physics Today* **45**, 26 (November 1992).

Hebard, A.F. Buckminsterfullerene. *Annual Review of Materials Science* **23**, 159–89 (1993).
Schön, J.H., Kloc, Ch. and Batlogg, B. Superconductivity at 52K in Hole-Doped C_{60}. *Nature* **400**, 549–52 (2000).

Scientific misconduct

Bumfield, G. *Nature* **419**, 419–20 (3 October 2002).
Durrani, M. *Physics World* **15**, 6–7 (November 2002).
Goodstein, J. *Physics World* **15**, 17–18 (November 2002).
Gwynne, P. *Physics World* **15**, 5 (June 2002).
Service, R.F. *Science* **298**, 30–1 (4 October 2002).

Chapter 11

Alexandrov, A.S. and Mott, N.F. High Temperature Superconductors and other Superfluids. Taylor and Francis (1994).
Allen, P.B. Is kinky conventional? *Nature* **412**, 494–5 (August 2 2001).
Anderson, P.W. Condensed Matter – The Continuous Revolution. *Physics World* **8**, 37–40 (December 1995).
Anderson, P.W. High Temperature Superconductivity Debate Heats Up. *Physics World* **9**, p. 16 (January 1996).
Anderson, P.W. and Ren, Y. The Normal State of High T_c Superconductors: A New Quantum Liquid The Los Alamos Symposium-1989 High Temperature Superconductivity Proceedings, ed. K.S. Bedell, D. Coffey, D.E. Meltzer, D. Pines and J.R. Schrieffer. Addison-Wesley Publishers. 3–34 (1990).
Anderson, P.W. The Theory of Superconductivity in the High-T_c Cuprates. Princeton University Press (1997).
Anderson, P.W. and Schrieffer, J.R. A Dialogue on the Theory of High T_c. *Physics Today* **44**, 54–61 (June 1991).
Annett, J.F. Unconventional Superconductivity. *Contemp. Phys.* **36**, 423–37 (1995).
Batlogg, B. and Varma, C. The Underdoped Phase of Cuprate Superconductors. *Physics World* **13**, 33–7 (February 2000).
Cox, D.L. and Maple, M.B. Electronic Pairing in Exotic Superconductors. *Physics Today* **48**, 32–40 (February 1995).
Fischer, O. Superconductivity and Magnetism in Chevrel Phases. "Earlier and Recent Aspects of Superconductivity" eds. Bednorz, J.G. and Müller, K.A.; Springer Series in Solid State Sciences **90**, Springer-Verlag, pp. 96–112 (1990).
Ford, P.J. Spin Glasses. *Contemporary Physics* **23**, 141–68 (1982).
Friedel, J. The high-T_c superconductors: a conservative view. *J. Phys. Condens. Matter* **1**, 7757–94 (1989).
Ginzburg, V.L. Once again about high-temperature superconductivity. *Contemporary Physics* **33**, 15–23 (1992).
Hussey, N.E. Low-Energy Quasiparticles in High-T_c Cuprates. *Adv. in Physics* **51**, 1685–1771 (2002)
Kirtley, J.R. and Tsuei, C.C. Probing High Temperature Superconductivity. *Scientific American* **275**, 50–5 (August 1996).
Lanzara, A., Bogdanov, P.V., Zhou, X.J., Kellar, S.A., Feng, D.L., Lu, E.D., Yoshida, T., Esaki, H., Fujimori, A., Kishio, K., Shimoyama, J.-L., Noda, T., Uchida, S., Hussein, Z. and Shen, Z.-X. Evidence for Ubuquitous Strong Electron–Phonon Coupling in High-Temperature Superconductors. *Nature* **412**, 510–14 (2 August 2001).
Mott, N.F. Is there an Explanation? *Nature* **327**,185–6 (1987).
Mott, N.F. High Temperature Superconductivity: The Spin Polaron Theory. *Contemp. Phys.* **30**, 373–84 (1990).

Mott, N.F. High Temperature Superconductivity Debate Heats Up. *Physics World* 9, p. 16 (January 1996).

Norman, M.R. and Pepin, C. The Electronic Nature of High Temperature Cuprate Superconductors. *Reports of Progress in Physics* 66, 1547–1610 (2003).

Ramakrishnan, T.V. High Temperature Superconductors: Facts and Theories. Critical Problems in Physics ed. V. Fitch, D.R. Marlow and M.A.E. Dementi. Princetown University Press, Princetown, New Jersey Chapter 5, 75–112 (1997).

Schofield, A. J. Non-Fermi Liquids. *Contemp. Phys.* 40, 95–115 (1999).

Steglish, F. Superconductivity of Strongly Correlated Electrons: Heavy-Fermion Systems. "Earlier and Recent Aspects of Superconductivity" eds Bednorz, J.G. and Müller, K.A.; Springer Series in Solid State Sciences 90, Springer-Verlag p. 306–25 (1990).

Tsuei, C.C. and Kirtley, J.R. Pairing Symmetry in Cuprate Superconductors. *Rev. Mod. Phys.* 72, 969–1016 (2000).

Symmetry

Pagels, H.R. The Cosmic Code – Quantum Physics as the Language of Nature. M. Joseph, London (1982).

Gribbin, J. In Search of SUSY – Supersymmetry, String and the Theory of Everything. Penguin Group (1998).

Epilogue

Crow, M. Technology Development in Japan and the United States: Lessons from the High-Temperature Superconductivity Race. *Science and Public Policy* 16, 322–44 (1989).

Feynman, R.P. Minority Report to the Space Shuttle Challenger Inquiry. The Pleasure of Finding Things Out – The Best Short Works of Richard P. Feynman. Chapter 7, pp. 151–69 Penguin Books (2001).

Maddox, J. Chemists Come in from the Cold. *Nature* 328, 663 (1987).

Pauling, L. Influence of Valence, Electronegativity, Atomic Radii and Crest-Trough Interaction with Phonons in the High-Temperature-Copper-Oxide Superconductors. *Phys. Rev. Letters* 59, 225–7 (1987).

Rowell, J.M. Superconductivity Research: A Different View. *Physics Today* 41, 36–42 (November 1988).

Rowell, J.M. Superconducting Electronics – The Next Decade "Proceedings of the 10th Anniversary HTS Workshop on Physics, Materials and Applications" ed. Batlogg, B., Chu, C.W., Gubser, D.U, and Müller, K.A. pp. 52–7, World Scientific Publishing Co. Singapore (1997).

Rowell, J.M. Some Reflections on Five Years of High Temperature Superconductivity. *Supercond. Sci. Technol.* 4, 692–5 (1991).

Rowell, J.M. Research Advances and Needs Related to HTSC Electronics. *Supercond. Sci. Technol.* 4, S51–8 (1991).

Rowell, J.M. Recommended Directions of Research and Development in Superconducting Electronics. *IEEE Transactions on Applied Superconductivity* 9, 2837–48 (1999).

Tanaka, S. Research on High-T_c Superconductivity in Japan. *Physics Today* 40, 53–7 (December 1987).

Subject Index

Index of Names[†]

[†]Names appearing only in the references have not been included